Standards Practice B[...]
For Home or School
Grade 4

INCLUDES:
- Home or School Practice
- Lesson Practice and Test Preparation
- English and Spanish School-Home Letters
- Getting Ready for Grade 5 Lessons

Place Value and Operations with Whole Numbers

Developing understanding and fluency with multi-digit multiplication, and developing understanding of dividing to find quotients involving multi-digit dividends

1 Place Value, Addition, and Subtraction to One Million

Domain Number and Operations in Base Ten

2 — Multiply by 1-Digit Numbers

Domains Operations and Algebraic Thinking
Number and Operations in Base Ten

3 — Multiply 2-Digit Numbers

Domains Operations and Algebraic Thinking
Number and Operations in Base Ten

© Houghton Mifflin Harcourt Publishing Company

Fractions and Decimals

Developing an understanding of fraction equivalence, addition and subtraction of fractions with like denominators, and multiplication of fractions by whole numbers

6 Fraction Equivalence and Comparison

Domain Number and Operations—Fractions

7 Add and Subtract Fractions

Domain Number and Operations—Fractions

8 Multiply Fractions by Whole Numbers

Domain Number and Operations—Fractions

9 Relate Fractions and Decimals

Domains Number and Operations—Fractions
Measurement and Data

© Houghton Mifflin Harcourt Publishing Company

Geometry, Measurement, and Data

Understanding that geometric figures can be analyzed and classified based on their properties, such as having parallel sides, perpendicular sides, particular angle measures, and symmetry

10 Two-Dimensional Figures

Domains Geometry
Operations and Algebraic Thinking

11 Angles

Domain Measurement and Data

© Houghton Mifflin Harcourt Publishing Company

12 Relative Sizes of Measurement Units

Domain Measurement and Data

13 Algebra: Perimeter and Area

Domain Measurement and Data

End-of-Year Resources

Getting Ready for Grade 5

These lessons review important skills and prepare you for Grade 5.

Table of Contents
Florida Lessons

School-Home Letter

Dear Family,

During the next few weeks, our math class will be learning how to use and represent whole numbers through the hundred thousands place. We will also be adding and subtracting multi-digit numbers.

You can expect to see homework that provides practice with naming numbers in different ways, as well as rounding and estimating greater numbers.

Here is a sample of how your child will be taught to write numbers in different forms.

Vocabulary

estimate A number that is close to the exact amount

expanded form A way to write numbers by showing the value of each digit

period Each group of three digits separated by commas in a multi-digit number

round To replace a number with another number that tells about how many or how much

standard form A way to write numbers using the digits 0–9 with each digit having a place value

word form A way to write numbers by using words

🔒 MODEL Place Value Through Hundred Thousands

This is how we will be writing numbers in different forms.

THOUSANDS			ONES		
Hundreds	Tens	Ones	Hundreds	Tens	Ones
2	8	1,	3	6	5

STANDARD FORM:

281,365

WORD FORM:

two hundred eighty-one thousand, three hundred sixty-five

EXPANDED FORM:

200,000 + 80,000 + 1,000 + 300 + 60 + 5

Tips

Rounding Greater Numbers

When rounding, first find the place to which you want to round. Then, look at the digit to the right. If the digit to the right is *less than* 5, the digit in the rounding place stays the same. If the digit is 5 *or greater*, the digit in the rounding place increases by 1. All the digits to the right of the rounding place change to zero.

Carta para la casa

Querida familia,

Durante las próximas semanas, en la clase de matemáticas aprenderemos cómo usar y representar números enteros hasta las centenas de millar. También vamos a sumar y restar números de varios dígitos.

Llevaré a la casa tareas que sirven para practicar diferentes maneras de expresar los números, además de redondear y estimar números mayores.

Este es un ejemplo de la manera como aprenderemos a expresar números de diferentes formas.

Vocabulario

estimación Un número que se aproxima a una cantidad exacta

forma desarrollada Una manera de escribir números que muestra el valor de cada dígito

periodo En un número de varios dígitos, cada grupo de tres dígitos separado por comas

redondear Reemplazar un número con otro que muestra una aproximación de cuánto o cuántos

forma estándar Una manera de escribir números usando los dígitos 0 a 9, en la que cada dígito tiene un valor posicional

en palabras Una manera de escribir números usando palabras

 MODELO Valor posicional hasta las centenas de millar

Así es como escribiremos números de diferentes formas.

MILLARES			UNIDADES		
Centenas	Decenas	Unidades	Centenas	Decenas	Unidades
2	8	1,	3	6	5

FORMA NORMAL:

281,365

EN PALABRAS:

doscientos ochenta y un mil, trescientos sesenta y cinco

FORMA DESARROLLADA:

200,000 + 80,000 + 1,000 + 300 + 60 + 5

Pistas

Redondear números grandes

Cuando se redondea, primero se halla el lugar al que se quiere redondear. Después, se debe mirar el dígito que está a la derecha. Si el dígito a la derecha es *menor que* 5, el dígito en el lugar del redondeo se queda igual. Si el dígito es 5 o *mayor*, el dígito en el lugar del redondeo aumenta en 1. Todos los dígitos a la derecha del lugar del redondeo cambian a cero.

Name _____

Model Place Value Relationships

Find the value of the underlined digit.

1. 6,0<u>3</u>5

2. 43,<u>7</u>82

3. 506,08<u>7</u>

4. 4<u>9</u>,254

_____ _____ _____ _____

5. 1<u>3</u>6,422

6. 673,<u>5</u>12

7. <u>8</u>14,295

8. 73<u>6</u>,144

_____ _____ _____ _____

Compare the values of the underlined digits.

9. 6,<u>3</u>00 and 5<u>3</u>0

The value of 3 in _____ is _____ times

the value of 3 in _____ .

10. <u>2</u>,783 and 7,<u>2</u>83

The value of 2 in _____ is _____ times

the value of 2 in _____ .

11. 3<u>4</u>,258 and <u>4</u>7,163

The value of 4 in _____ is _____ times

the value of 4 in _____ .

12. 503,49<u>7</u> and 26,4<u>7</u>5

The value of 7 in _____ is _____ times

the value of 7 in _____ .

Problem Solving REAL WORLD

Use the table for 13–14.

13. What is the value of the digit 9 in the attendance at the Chargers vs. Titans game?

_____

14. The attendance at which game has a 7 in the ten thousands place?

_____

Football Game Attendance	
Game	**Attendance**
Chargers vs. Titans	69,143
Ravens vs. Panthers	73,021
Patriots vs. Colts	68,756

Lesson Check

1. During one season, a total of 453,193 people attended a baseball team's games. What is the value of the digit 5 in the number of people?

   Ⓐ   500

   Ⓑ   5,000

   Ⓒ   50,000

   Ⓓ 500,000

2. Hal forgot the number of people at the basketball game. He does remember that the number had a 3 in the tens place. Which number could Hal be thinking of?

   Ⓐ 7,321

   Ⓑ 3,172

   Ⓒ 2,713

   Ⓓ 1,237

Spiral Review

3. Hot dog buns come in packages of 8. For the school picnic, Mr. Spencer bought 30 packages of hot dog buns. How many hot dog buns did he buy? (Grade 3)

   Ⓐ   24

   Ⓑ   38

   Ⓒ 110

   Ⓓ 240

4. There are 8 students on the minibus. Five of the students are boys. What fraction of the students are boys? (Grade 3)

   Ⓐ $\frac{3}{8}$

   Ⓑ $\frac{5}{8}$

   Ⓒ $\frac{5}{5}$

   Ⓓ $\frac{8}{8}$

5. The clock below shows the time when Amber leaves home for school. At what time does Amber leave home? (Grade 3)

   Ⓐ 2:41   Ⓒ 8:10

   Ⓑ 8:02   Ⓓ 8:20

6. Jeremy drew a polygon with four right angles and four sides with the same length.

   What kind of polygon did Jeremy draw? (Grade 3)

   Ⓐ hexagon

   Ⓑ square

   Ⓒ trapezoid

   Ⓓ triangle

Name _____

Read and Write Numbers

Read and write the number in two other forms.

1. six hundred ninety-two thousand, four

standard form:
692,004;
expanded form:
600,000 +
90,000 +
2,000 + 4

2. 314,207

_____

_____

_____

_____

3. 600,000 + 80,000 + 10

_____

_____

_____

_____

Use the number 913,256.

4. Write the name of the period that has the digits 913.

_____

5. Write the digit in the ten thousands place.

_____

6. Write the value of the digit 9.

_____

Problem Solving  REAL WORLD

Use the table for 7 and 8.

Population in 2008

State	Population
Alaska	686,293
South Dakota	804,194
Wyoming	532,668

7. Which state had a population of eight hundred four thousand, one hundred ninety-four?

_____

8. What is the value of the digit 8 in Alaska's population?

_____

Lesson Check

1. Based on a 2008 study, children 6–11 years old spend sixty-nine thousand, one hundred eight minutes a year watching television. What is this number written in standard form?

   (A) 6,918

   (B) 69,108

   (C) 69,180

   (D) 690,108

2. What is the value of the digit 4 in the number 84,230?

   (A)   4

   (B)   400

   (C) 4,000

   (D) 40,000

Spiral Review

3. An ant has 6 legs. How many legs do 8 ants have in all? (Grade 3)

   (A) 14

   (B) 40

   (C) 45

   (D) 48

4. Latricia's vacation is in 4 weeks. There are 7 days in a week. How many days is it until Latricia's vacation? (Grade 3)

   (A) 9 days

   (B) 11 days

   (C) 20 days

   (D) 28 days

5. Marta collected 363 cans. Diego collected 295 cans. How many cans did Marta and Diego collect in all? (Grade 3)

   (A) 668

   (B) 658

   (C) 568

   (D) 178

6. The city Tim lives in has 106,534 people. What is the value of the 6 in 106,534? (Lesson 1.1)

   (A) 6,000

   (B)   600

   (C)   60

   (D)   6

Name _____

Compare and Order Numbers

Compare. Write <, >, or =.

1. 3,273 $\boxed{<}$ 3,279

2. $1,323 $\bigcirc$ $1,400

3. 52,692 $\bigcirc$ 52,692

4. $413,005 $\bigcirc$ $62,910

5. 382,144 $\bigcirc$ 382,144

6. 157,932 $\bigcirc$ 200,013

7. 401,322 $\bigcirc$ 410,322

8. 989,063 $\bigcirc$ 980,639

9. 258,766 $\bigcirc$ 258,596

Order from least to greatest.

10. 23,710; 23,751; 23,715

_____

11. 52,701; 54,025; 5,206

_____

12. 465,321; 456,321; 456,231

_____

13. $330,820; $329,854; $303,962

_____

Problem Solving REAL WORLD

14. An online newspaper had 350,080 visitors in October, 350,489 visitors in November, and 305,939 visitors in December. What is the order of the months from greatest to least number of visitors?

_____

_____

15. The total land area in square miles of each of three states is shown below.
    Colorado: 103,718
    New Mexico: 121,356
    Arizona: 113,635
    What is the order of the states from least to greatest total land area?

_____

_____

Lesson Check

1. At the yearly fund-raising drive, the nonprofit company's goal was to raise $55,500 each day. After three days, it had raised $55,053; $56,482; and $55,593. Which amount was less than the daily goal?

   Ⓐ $55,500   Ⓒ $55,593

   Ⓑ $55,053   Ⓓ $56,482

2. Which of the following lists of numbers is in order from greatest to least?

   Ⓐ 60,343; 60,433; 63,043

   Ⓑ 83,673; 86,733; 86,373

   Ⓒ 90,543; 90,048; 93,405

   Ⓓ 20,433; 20,343; 20,043

Spiral Review

3. Jess is comparing fractions. Which fraction is greater than $\frac{5}{6}$? **(Grade 3)**

   Ⓐ $\frac{7}{8}$

   Ⓑ $\frac{4}{5}$

   Ⓒ $\frac{3}{4}$

   Ⓓ $\frac{2}{3}$

4. What is the perimeter of the rectangle below? **(Grade 3)**

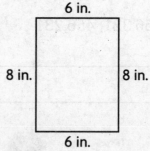

6 in.

8 in.   8 in.

6 in.

   Ⓐ 14 inches

   Ⓑ 26 inches

   Ⓒ 28 inches

   Ⓓ 48 inches

5. A website had 826,140 hits last month. What is the value of the 8 in 826,140?

   (Lesson 1.1)

   Ⓐ   800

   Ⓑ   8,000

   Ⓒ   80,000

   Ⓓ   800,000

6. Which is 680,705 written in expanded form? **(Lesson 1.2)**

   Ⓐ 680 + 705

   Ⓑ 68,000 + 700 + 5

   Ⓒ 600,000 + 8,000 + 700 + 5

   Ⓓ 600,000 + 80,000 + 700 + 5

Name _____

Round Numbers

Round to the place value of the underlined digit.

1. 8<u>6</u>2,840

8<u>6</u>2,840 _____**860,000**_____
↑
less than 5

2. 123,<u>4</u>99

_____

3. <u>5</u>52,945

_____

- Look at the digit to the right. If the digit to the right is *less than 5*, the digit in the rounding place stays the same.

- Change all the digits to the right of the rounding place to zero.

4. 38<u>9</u>,422

_____

5. <u>2</u>09,767

_____

6. 191,<u>3</u>06

_____

7. <u>6</u>6,098

_____

8. 73,<u>5</u>90

_____

9. <u>1</u>49,903

_____

10. 684,<u>3</u>03

_____

11. 499,<u>5</u>53

_____

Problem Solving  REAL WORLD

Use the table for 12–13.

12. Find the height of Mt. Whitney in the table. Round the height to the nearest thousand feet.

_____ feet

13. What is the height of Mt. Bona rounded to the nearest ten thousand feet?

_____ feet

Mountain Heights		
Name	**State**	**Height (feet)**
Mt. Bona	Alaska	16,500
Mt. Whitney	California	14,494

Lesson Check

1. Which number is 247,039 rounded to the nearest thousand?

   Ⓐ 200,000   Ⓒ 247,000

   Ⓑ 250,000   Ⓓ 7,000

2. To the nearest ten thousand, the population of Vermont was estimated to be about 620,000 in 2008. Which might have been the exact population of Vermont in 2008?

   Ⓐ 626,013   Ⓒ 614,995

   Ⓑ 621,270   Ⓓ 609,964

Spiral Review

3. Which symbol makes the following number sentence true? **(Lesson 1.3)**

   $546,322 ◯ $540,997

   Ⓐ <

   Ⓑ >

   Ⓒ =

   Ⓓ +

4. Pittsburgh International Airport had approximately 714,587 passengers in August 2009. Which number is greater than 714,587? **(Lesson 1.3)**

   Ⓐ 714,578

   Ⓑ 704,988

   Ⓒ 714,601

   Ⓓ 714,099

5. June made a design with 6 equal tiles. One tile is yellow, 2 tiles are blue, and 3 tiles are purple. What fraction of the tiles are yellow or purple? **(Grade 3)**

   Ⓐ $\frac{1}{6}$

   Ⓑ $\frac{2}{6}$

   Ⓒ $\frac{3}{6}$

   Ⓓ $\frac{4}{6}$

6. The fourth grade collected 40,583 cans and plastic bottles. Which of the following shows that number in word form? **(Lesson 1.2)**

   Ⓐ forty thousand, five hundred eighty

   Ⓑ forty thousand, five hundred eighty-three

   Ⓒ four thousand, five hundred eighty-three

   Ⓓ four hundred thousand, five hundred eighty-three

Name _____

Rename Numbers

Rename the number. Use the place-value chart to help.

1. 760 hundreds = __**76,000**__

THOUSANDS			ONES		
Hundreds	Tens	Ones	Hundreds	Tens	Ones
	7	6,	0	0	0

2. 805 tens = _____

THOUSANDS			ONES		
Hundreds	Tens	Ones	Hundreds	Tens	Ones

3. 24 ten thousands = _____

THOUSANDS			ONES		
Hundreds	Tens	Ones	Hundreds	Tens	Ones

Rename the number.

4. 720 = _____ tens

5. 4 thousands 7 hundreds = 47 _____

6. 25,600 = _____ hundreds

7. 204 thousands = _____

Problem Solving REAL WORLD

8. For the fair, the organizers ordered 32 rolls of tickets. Each roll of tickets has 100 tickets. How many tickets were ordered in all?

_____

9. An apple orchard sells apples in bags of 10. The orchard sold a total of 2,430 apples one day. How many bags of apples was this?

_____

Lesson Check

1. A dime has the same value as 10 pennies. Marley brought 290 pennies to the bank. How many dimes did Marley get?

   (A) 29
   (B) 290
   (C) 2,900
   (D) 29,000

2. A citrus grower ships grapefruit in boxes of 10. One season, the grower shipped 20,400 boxes of grapefruit. How many grapefruit were shipped?

   (A) 204
   (B) 2,040
   (C) 20,400
   (D) 204,000

Spiral Review

3. There were 2,605 people at the basketball game. A reporter rounded this number to the nearest hundred for a newspaper article. What number did the reporter use? (Lesson 1.4)

   (A) 2,600   (C) 2,700
   (B) 2,610   (D) 3,000

4. To get to Level 3 in a game, a player must score 14,175 points. Ann scores 14,205 points, Ben scores 14,089 points, and Chuck scores 10,463 points. Which score is greater than the Level 3 score? (Lesson 1.3)

   (A) 14,205   (C) 14,089
   (B) 14,175   (D) 10,463

5. Henry counted 350 lockers in his school. Hayley counted 403 lockers in her school. Which statement is true? (Lesson 1.1)

   (A) The 3 in 350 is 10 times the value of the 3 in 403.
   (B) The 3 in 350 is 100 times the value of the 3 in 403.
   (C) The 3 in 403 is 10 times the value of the 3 in 350.
   (D) The 3 in 403 is 100 times the value of the 3 in 350.

6. There are 4 muffins on each plate. There are 0 plates of lemon muffins. How many lemon muffins are there? (Grade 3)

   (A) 4
   (B) 2
   (C) 1
   (D) 0

Name _____

Add Whole Numbers

Estimate. Then find the sum.

1. Estimate: **90,000**

$$
\begin{array}{r}
\overset{1\,1}{63{,}824} \rightarrow 60{,}000 \\
+\ 29{,}452 \rightarrow +\ 30{,}000 \\
\hline
93{,}276 \qquad 90{,}000
\end{array}
$$

2. Estimate: _____

$$
\begin{array}{r}
73{,}404 \\
+\ 27{,}865 \\
\hline
\end{array}
$$

3. Estimate: _____

$$
\begin{array}{r}
403{,}446 \\
+\ 396{,}755 \\
\hline
\end{array}
$$

4. Estimate: _____

$$
\begin{array}{r}
137{,}638 \\
+\ 52{,}091 \\
\hline
\end{array}
$$

5. Estimate: _____

$$
\begin{array}{r}
200{,}629 \\
+\ 28{,}542 \\
\hline
\end{array}
$$

6. Estimate: _____

$$
\begin{array}{r}
212{,}514 \\
+\ 396{,}705 \\
\hline
\end{array}
$$

7. Estimate: _____

$$
\begin{array}{r}
324{,}867 \\
+\ 6{,}233 \\
\hline
\end{array}
$$

8. Estimate: _____

$$
\begin{array}{r}
462{,}809 \\
+\ 256{,}738 \\
\hline
\end{array}
$$

9. Estimate: _____

$$
\begin{array}{r}
624{,}836 \\
+\ 282{,}189 \\
\hline
\end{array}
$$

Problem Solving REAL WORLD

Use the table for 10–12.

10. Beth and Cade were on one team. What was their total score?

    _____

11. Dillan and Elaine were on the other team. What was their total score?

    _____

12. Which team scored the most points?

    _____

Individual Game Scores	
Student	**Score**
Beth	251,567
Cade	155,935
Dillan	188,983
Elaine	220,945

Lesson Check

1. The coastline of the United States is 12,383 miles long. Canada's coastline is 113,211 miles longer than the coastline of the United States. How long is the coastline of Canada?

   (A) 100,828 miles

   (B) 115,594 miles

   (C) 125,594 miles

   (D) 237,041 miles

2. Germany is the seventh largest European country and is slightly smaller by area than Montana. Germany has a land area of 134,835 square miles and a water area of 3,011 square miles. What is the total area of Germany?

   (A) 7,846 square miles

   (B) 131,824 square miles

   (C) 137,846 square miles

   (D) 435,935 square miles

Spiral Review

3. In an election, about 500,000 people voted in all. Which number could be the exact number of people who voted in the election? (Lesson 1.4)

   (A) 429,455

   (B) 441,689

   (C) 533,736

   (D) 550,198

4. In 2007, Pennsylvania had approximately 121,580 miles of public roads. What is 121,580 rounded to the nearest thousand? (Lesson 1.4)

   (A) 100,000

   (B) 120,000

   (C) 121,000

   (D) 122,000

5. Which of the following lists of numbers is in order from greatest to least? (Lesson 1.3)

   (A) 33,093; 33,903; 33,309

   (B) 42,539; 24,995; 43,539

   (C) 682,131; 628,000; 682,129

   (D) 749,340; 740,999; 740,256

6. Which symbol makes the following statement true? (Lesson 1.3)

   $413,115 ◯ $431,511

   (A) <

   (B) >

   (C) =

   (D) +

Name _____

Subtract Whole Numbers

Estimate. Then find the difference.

1. Estimate: __600,000__

```
      9
   7 10 15 6 13
  7 8̶0̶,5̶7̶3̶
- 2 2 9,6 1 5
  5 5 0,9 5 8
```

Think: 780,573 rounds to 800,000.

229,615 rounds to 200,000.

So an estimate is 800,000 − 200,000 = 600,000.

2. Estimate: _____

```
  428,731
- 175,842
```

3. Estimate: _____

```
  920,026
- 535,722
```

4. Estimate: _____

```
  253,495
-  48,617
```

Subtract. Add to check.

5. 735,249 − 575,388

6. 512,724 − 96,473

7. 600,000 − 145,782

Problem Solving REAL WORLD

Use the table for 8 and 9.

8. How many more people attended the Magic's games than attended the Pacers' games?

_____

9. How many fewer people attended the Pacers' games than attended the Clippers' games?

_____

Season Attendance for Three NBA Teams	
Team	Attendance
Indiana Pacers	582,295
Orlando Magic	715,901
Los Angeles Clippers	670,063

Lesson Check

1. This year, a farm planted 400,000 corn stalks. Last year, the farm planted 275,650 corn stalks. How many more corn stalks did the farm plant this year than last year?

    (A) 124,350

    (B) 125,450

    (C) 235,450

    (D) 275,650

2. One machine can make 138,800 small paper clips in one day. Another machine can make 84,250 large paper clips in one day. How many more small paper clips than large paper clips are made by the two machines in one day?

    (A) 44,550

    (B) 54,550

    (C) 54,650

    (D) 154,650

Spiral Review

3. In three baseball games over a weekend, 125,429 people came to watch. The next weekend, 86,353 came to watch the games. How many people in all watched the six baseball games? (Lesson 1.6)

    (A) 201,782

    (B) 211,772

    (C) 211,782

    (D) 211,882

4. Kevin read the number "two hundred seven thousand, forty-eight" in a book. What is this number in standard form? (Lesson 1.2)

    (A) 27,048

    (B) 27,480

    (C) 207,048

    (D) 207,480

5. A museum had 275,608 visitors last year. What is this number rounded to the nearest thousand? (Lesson 1.4)

    (A) 275,600

    (B) 276,000

    (C) 280,000

    (D) 300,000

6. At the Millville Theater, a play ran for several weeks. In all, 28,175 people saw the play. What is the value of the digit 8 in 28,175? (Lesson 1.1)

    (A) 8

    (B) 800

    (C) 8,000

    (D) 80,000

Name _____

Problem Solving • Comparison Problems with Addition and Subtraction

Use the information in the table for 1–3.

1. How many square miles larger is the surface area of Lake Huron than the surface area of Lake Erie?

   Think: How can a bar model help represent the problem? What equation can be written?

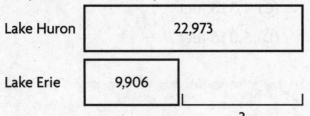

   | Lake Huron | 22,973 |

   | Lake Erie | 9,906 |

   ?

   $22,973 - 9,906 =$ __13,067__ square miles

   13,067 square miles

Surface Area of the Great Lakes	
Lake	Surface Area (in square miles)
Lake Superior	31,700
Lake Michigan	22,278
Lake Huron	22,973
Lake Erie	9,906
Lake Ontario	7,340

2. Which lake has a surface area that is 14,938 square miles greater than the surface area of Lake Ontario? Draw a model and write a number sentence to solve the problem.

   _____

3. Lake Victoria has the largest surface area of all lakes in Africa. Its surface area is 26,828 square miles. How much larger is the surface area of Lake Superior than that of Lake Victoria?

   _____

4. At 840,000 square miles, Greenland is the largest island in the world. The second-largest island is New Guinea, at 306,000 square miles. How much larger is Greenland than New Guinea?

   _____

Lesson Check

1. The Mariana Trench in the Pacific Ocean is about 36,201 feet deep. The Puerto Rico Trench in the Atlantic Ocean is about 27,493 feet deep. Based on these data, how many feet deeper is the Mariana Trench than the Puerto Rico Trench?

   (A) 8,708 feet   (C) 9,808 feet

   (B) 9,718 feet   (D) 63,694 feet

2. At 1,932 feet, Crater Lake, Oregon, is the deepest lake in the United States. The world's deepest lake, Lake Baykal in Russia, is 3,383 feet deeper. How deep is Lake Baykal?

   (A) 3,383 feet

   (B) 4,215 feet

   (C) 4,315 feet

   (D) 5,315 feet

Spiral Review

3. Which of the following amounts is greater than $832,458? (Lesson 1.3)

   (A) $82,845

   (B) $832,458

   (C) $823,845

   (D) $832,485

4. A stadium in Pennsylvania seats 107,282 people. A stadium in Arizona seats 71,706 people. Based on these facts, how many more people does the stadium in Pennsylvania seat than the stadium in Arizona? (Lesson 1.7)

   (A) 35,576   (C) 36,576

   (B) 35,586   (D) 178,988

5. Which of the following numbers is 399,713 rounded to the place value of the underlined digit? (Lesson 1.4)

   (A) 390,000

   (B) 398,000

   (C) 399,800

   (D) 400,000

6. About 400,000 people visited an art museum in December. Which number could be the exact number of people who visited the art museum? (Lesson 1.4)

   (A) 478,051

   (B) 452,223

   (C) 352,483

   (D) 348,998

Name _____

Chapter 1 Extra Practice

Lesson 1.1

Find the value of the underlined digit.

1. 6,4<u>9</u>3   2. <u>1</u>6,403   3. 725,<u>3</u>60   4. <u>9</u>52,635

_____   _____   _____   _____

Compare the values of the underlined digits in <u>4</u>6,395 and 1<u>4</u>,906.

5. The value of 4 in _____ is ____ times

the value of 4 in _____.

Lesson 1.2

Read and write the number in two other forms.

1. 304,001

_____

_____

_____

2. two hundred eight thousand, five hundred sixty-one

_____

_____

_____

Use the number 751,486.

3. Write the name of the period that has the digits 486.

_____

4. Write the name of the period that has the digits 751.

_____

5. Write the digit in the thousands place.

_____

6. Write the value of the digit 5.

_____

Lesson 1.3

Compare. Write <, >, or =.

1. 6,930 ◯ 7,023   2. 98,903 ◯ 98,930   3. 549,295 ◯ 547,364

Order from least to greatest.

4. $26,940; $25,949; $26,490

_____

5. 634,943; 639,443; 589,932

_____

Lesson 1.4

Round to the place value of the underlined digit.

1. 286,476
2. 289,342
3. 245,001
4. 183,002

_____   _____   _____   _____

Lesson 1.5

Rename the number.

1. 82 thousands = _____

2. 600,000 = _____ ten thousands

3. 9,200 = _____ hundreds

4. 8 ten thousands 4 hundreds = _____

Lesson 1.6

Estimate. Then find the sum.

1. Estimate: _____

$$94,903 + 49,995$$

2. Estimate: _____

$$420,983 + 39,932$$

3. Estimate: _____

$$540,943 + 382,093$$

Lesson 1.7

Estimate. Then find the difference.

1. Estimate: _____

$$25,953 - 9,745$$

2. Estimate: _____

$$740,758 - 263,043$$

3. Estimate: _____

$$807,632 - 592,339$$

Lesson 1.8

1. The attendance for the first game of the football season was 93,584. The attendance for the second game was 104,227. How many more people attended the second game than the first game?

2. Abby and Lee sold raffle tickets to raise money for a new playground. Abby sold 1,052 tickets. Lee sold 379 more tickets than Abby. How many tickets did Lee sell?

_____   _____

School-Home Letter

© Houghton Mifflin Harcourt Publishing Company

Dear Family,

During the next few weeks, our math class will be learning about multiplying by 1-digit whole numbers. We will investigate strategies for multiplying 2-, 3-, and 4-digit numbers by the numbers 2–9.

You can expect to see homework that provides practice with multiplication by 1-digit numbers.

Here is a sample of how your child will be taught to multiply by a 1-digit number.

Vocabulary

Distributive Property The property that states that multiplying a sum by a number is the same as multiplying each addend by the number and then adding the products

partial products A method of multiplying in which the ones, tens, hundreds, and so on are multiplied separately and then the products are added together

🔑 MODEL Multiply by a 1-Digit Number

This is one way we will be multiplying by 1-digit numbers.

STEP 1

Multiply the tens. Record.

$$\begin{array}{r} 26 \\ \times\ 3 \\ \hline 60 \end{array} \leftarrow 3 \times 2 \text{ tens} = 6 \text{ tens}$$

STEP 2

Multiply the ones. Record.

$$\begin{array}{r} 26 \\ \times\ 3 \\ \hline 60 \\ 18 \end{array} \leftarrow 3 \times 6 \text{ ones} = 18 \text{ ones}$$

STEP 3

Add the partial products.

$$\begin{array}{r} 26 \\ \times\ 3 \\ \hline 60 \\ +\ 18 \\ \hline 78 \end{array}$$

Tips

Estimating to Check Multiplication

When estimation is used to check that a multiplication answer is reasonable, usually the greater factor is rounded to a multiple of 10 that has only one non-zero digit. Then mental math can be used to recall the basic fact product, and patterns can be used to determine the correct number of zeros in the estimate.

Carta para la casa

Querida familia,

Durante las próximas semanas, en la clase de matemáticas aprenderemos a multiplicar números enteros de un dígito. Investigaremos estrategias para multiplicar números de 2, 3 y 4 dígitos por números del 2 al 9.

Llevaré a la casa tareas para practicar la multiplicación de números de 1 dígito.

Este es un ejemplo de la manera como aprenderemos a multiplicar por un número de 1 dígito.

🔒 MODELO Multiplicar por un número de 1 dígito

Esta es una manera en la que multiplicaremos por un número de 1 dígito.

PASO 1

Multiplica las decenas. Anota.

$$
\begin{array}{r}
26 \\
\times\ 3 \\
\hline
60
\end{array}
$$
← 3 × 2 decenas = 6 decenas

PASO 2

Multiplica las unidades. Anota.

$$
\begin{array}{r}
26 \\
\times\ 3 \\
\hline
60 \\
18
\end{array}
$$
← 3 × 6 unidades = 18 unidades

PASO 3

Suma los productos parciales.

$$
\begin{array}{r}
26 \\
\times\ 3 \\
\hline
60 \\
+\ 18 \\
\hline
78
\end{array}
$$

Pistas

Estimar para revisar la multiplicación

Cuando se usa la estimación para revisar que la respuesta de una multiplicación es razonable, el factor se suele redondear al múltiplo de 10 que tiene un solo dígito distinto a cero. Después se puede usar el cálculo mental para recordar el producto básico de la operación, y se pueden usar patrones para determinar la cantidad correcta de ceros de la estimación.

Name _____

Multiplication Comparisons

Write a comparison sentence.

1. $6 \times 3 = 18$

   ___6___ times as many as ___3___ is ___18___.

2. $63 = 7 \times 9$

   _____ is _____ times as many as _____.

3. $5 \times 4 = 20$

   _____ times as many as _____ is _____.

4. $48 = 8 \times 6$

   _____ is _____ times as many as _____.

Write an equation.

5. 2 times as many as 8 is 16.

   _____

6. 42 is 6 times as many as 7.

   _____

7. 3 times as many as 5 is 15.

   _____

8. 36 is 9 times as many as 4.

   _____

9. 72 is 8 times as many as 9.

   _____

10. 5 times as many as 6 is 30.

    _____

Problem Solving  REAL WORLD

11. Alan is 14 years old. This is twice as old as his brother James is. How old is James?

    _____

12. There are 27 campers. This is nine times as many as the number of counselors. How many counselors are there?

    _____

Lesson Check

1. Which equation best represents the comparison sentence?

    24 is 4 times as many as 6.

    (A) $24 \times 4 = 6$

    (B) $24 = 4 \times 6$

    (C) $24 = 4 + 6$

    (D) $4 + 6 = 24$

2. Which comparison sentence best represents the equation?

    $5 \times 9 = 45$

    (A) 5 more than 9 is 45.

    (B) 9 is 5 times as many as 45.

    (C) 5 is 9 times as many as 45.

    (D) 45 is 5 times as many as 9.

Spiral Review

3. Which of the following statements correctly compares the numbers? (Lesson 1.3)

    (A) $273,915 > 274,951$

    (B) $134,605 < 143,605$

    (C) $529,058 > 530,037$

    (D) $452,731 > 452,819$

4. What is the standard form for $200,000 + 80,000 + 700 + 6$? (Lesson 1.2)

    (A)   2,876

    (B)   28,706

    (C) 208,706

    (D) 280,706

5. Sean and Leah are playing a computer game. Sean scored 72,491 points. Leah scored 19,326 points more than Sean. How many points did Leah score? (Lesson 1.6)

    (A) 53,615

    (B) 91,717

    (C) 91,815

    (D) 91,817

6. A baseball stadium has 38,496 seats. Rounded to the nearest thousand, how many seats is this? (Lesson 1.4)

    (A) 38,000

    (B) 38,500

    (C) 39,000

    (D) 40,000

Name _____

Comparison Problems

Draw a model. Write an equation and solve.

1. Stacey made a necklace using 4 times as many blue beads as red beads. She used a total of 40 beads. How many blue beads did Stacey use?

Think: Stacey used a total of 40 beads. Let n represent the number of red beads.

blue | n | n | n | n | } 40

red | n |

$5 \times n = 40; 5 \times 8 = 40;$

$4 \times 8 = 32$ blue beads

2. At the zoo, there were 3 times as many monkeys as lions. Tom counted a total of 24 monkeys and lions. How many monkeys were there?

_____

_____

3. Fred's frog jumped 7 times as far as Al's frog. The two frogs jumped a total of 56 inches. How far did Fred's frog jump?

4. Sheila has 5 times as many markers as Dave. Together, they have 18 markers. How many markers does Sheila have?

_____

_____

Problem Solving REAL WORLD

5. Rafael counted a total of 40 white cars and yellow cars. There were 9 times as many white cars as yellow cars. How many white cars did Rafael count?

6. Sue scored a total of 35 points in two games. She scored 6 times as many points in the second game as in the first. How many more points did she score in the second game?

_____

_____

© Houghton Mifflin Harcourt Publishing Company

Lesson Check

1. Sari has 3 times as many pencil erasers as Sam. Together, they have 28 erasers. How many erasers does Sari have?

   (A) 7

   (B) 14

   (C) 18

   (D) 21

2. In Sean's fish tank, there are 6 times as many goldfish as guppies. There are a total of 21 fish in the tank. How many more goldfish are there than guppies?

   (A) 5

   (B) 12

   (C) 15

   (D) 18

Spiral Review

3. Barbara has 9 stuffed animals. Trish has 3 times as many stuffed animals as Barbara. How many stuffed animals does Trish have? **(Lesson 2.1)**

   (A) 3

   (B) 12

   (C) 24

   (D) 27

4. There are 104 students in the fourth grade at Allison's school. One day, 15 fourth-graders were absent. How many fourth-graders were at school that day? **(Lesson 1.7)**

   (A) 89

   (B) 91

   (C) 99

   (D) 119

5. Joshua has 112 rocks. Jose has 98 rocks. Albert has 107 rocks. What is the correct order of the boys from the least to the greatest number of rocks owned? **(Lesson 1.3)**

   (A) Jose, Albert, Joshua

   (B) Jose, Joshua, Albert

   (C) Albert, Jose, Joshua

   (D) Joshua, Albert, Jose

6. Alicia has 32 stickers. This is 4 times as many stickers as Benita has. How many stickers does Benita have? **(Lesson 2.1)**

   (A) 6

   (B) 8

   (C) 9

   (D) 28

Name _____

Multiply Tens, Hundreds, and Thousands

Find the product.

1. $4 \times 7,000 =$ __28,000__

   Think: $4 \times 7 = 28$
   So, $4 \times 7,000 = 28,000$

2. $9 \times 60 =$ _____

3. $8 \times 200 =$ _____

4. $5 \times 6,000 =$ _____

5. $7 \times 800 =$ _____

6. $8 \times 90 =$ _____

7. $6 \times 3,000 =$ _____

8. $3 \times 8,000 =$ _____

9. $5 \times 500 =$ _____

10. $9 \times 4,000 =$ _____

11. $7 \times 7,000 =$ _____

12. $3 \times 40 =$ _____

13. $4 \times 5,000 =$ _____

14. $2 \times 9,000 =$ _____

Problem Solving REAL WORLD

15. A bank teller has 7 rolls of coins. Each roll has 40 coins. How many coins does the bank teller have?

    _____

16. Theo buys 5 packages of paper. There are 500 sheets of paper in each package. How many sheets of paper does Theo buy?

    _____

Lesson Check

1. A plane is traveling at a speed of 400 miles per hour. How far will the plane travel in 5 hours?

   (A)   200 miles

   (B)   2,000 miles

   (C) 20,000 miles

   (D) 200,000 miles

2. One week, a clothing factory made 2,000 shirts in each of 6 different colors. How many shirts did the factory make in all?

   (A)   2,000

   (B)   12,000

   (C) 120,000

   (D) 200,000

Spiral Review

3. Which comparison sentence best represents the equation? (Lesson 2.1)

   $$6 \times 7 = 42$$

   (A) 7 is 6 times as many as 42.

   (B) 6 is 7 times as many as 42.

   (C) 42 is 6 times as many as 7.

   (D) 6 more than 7 is 42.

4. The population of Middleton is six thousand, fifty-four people. Which of the following shows this number written in standard form? (Lesson 1.2)

   (A)   654

   (B) 6,054

   (C) 6,504

   (D) 6,540

5. In an election for mayor, 85,034 people voted for Carl Green and 67,952 people voted for Maria Lewis. By how many votes did Carl Green win the election? (Lesson 1.7)

   (A) 17,082

   (B) 17,182

   (C) 22,922

   (D) 152,986

6. Meredith picked 4 times as many green peppers as red peppers. If she picked a total of 20 peppers, how many green peppers did she pick? (Lesson 2.2)

   (A) 4

   (B) 5

   (C) 16

   (D) 24

Name _____

Estimate Products

Estimate the product by rounding.

1. 4×472

$$4 \times 472$$
↓
$$4 \times 500$$
2,000

2. $2 \times 6,254$

3. 9×54

4. $5 \times 5,503$

_____

5. 3×832

6. 6×98

7. $8 \times 3,250$

8. 7×777

_____

Find two numbers the exact answer is between.

9. 3×567

10. $6 \times 7,381$

11. 4×94

12. 8×684

_____

Problem Solving

13. Isaac drinks 8 glasses of water each day. He says he will drink 2,920 glasses of water in a year that has 365 days. Is the exact answer reasonable? **Explain.**

_____

_____

_____

14. Most Americans throw away about 1,365 pounds of trash each year. Is it reasonable to estimate that Americans throw away over 10,000 pounds of trash in 5 years? **Explain.**

_____

_____

_____

Lesson Check

1. A theater has 4,650 seats. If the theater sells all the tickets for each of its 5 shows, about how many tickets will the theater sell in all?

   Ⓐ 2,500   Ⓒ 25,000

   Ⓑ 10,000   Ⓓ 30,000

2. Washington Elementary has 4,358 students. Jefferson High School has 3 times as many students as Washington Elementary. About how many students does Jefferson High School have?

   Ⓐ 16,000   Ⓒ 10,000

   Ⓑ 12,000   Ⓓ 1,200

Spiral Review

3. Diego has 4 times as many autographed baseballs as Melanie has. Diego has 24 autographed baseballs. How many autographed baseballs does Melanie have? (Lesson 2.1)

   Ⓐ 28

   Ⓑ 20

   Ⓒ 8

   Ⓓ 6

4. Mr. Turkowski bought 4 boxes of envelopes at the office supply store. Each box has 500 envelopes. How many envelopes did Mr. Turkowski buy? (Lesson 2.3)

   Ⓐ 200

   Ⓑ 504

   Ⓒ 2,000

   Ⓓ 20,000

5. Pennsylvania has a land area of 44,816 square miles. Which of the following shows the land area of Pennsylvania rounded to the nearest hundred? (Lesson 1.4)

   Ⓐ 44,000 square miles

   Ⓑ 44,800 square miles

   Ⓒ 44,900 square miles

   Ⓓ 45,000 square miles

6. The table shows the types of DVDs customers rented from Sunshine Movie Rentals last year.

Movie Rentals	
Type	Number Rented
Comedy	6,720
Drama	4,032
Action	5,540

   How many comedy and action movies were rented in all last year? (Lesson 1.6)

   Ⓐ 13,620   Ⓒ 12,260

   Ⓑ 13,000   Ⓓ 10,752

Name _____

Multiply Using the Distributive Property

Model the product on the grid. Record the product.

1. $4 \times 19 = \underline{76}$

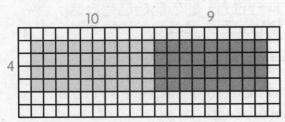

$4 \times 10 = 40$ and $4 \times 9 = 36$

$40 + 36 = 76$

2. $5 \times 13 = \underline{\hspace{1cm}}$

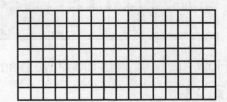

Find the product.

3. $4 \times 14 = \underline{\hspace{1cm}}$

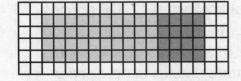

4. $3 \times 17 = \underline{\hspace{1cm}}$

5. $6 \times 15 = \underline{\hspace{1cm}}$

Problem Solving 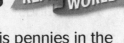 REAL WORLD

6. Michael arranged his pennies in the following display:

How many pennies does Michael have in all?

7. A farmer has an apple orchard with the trees arranged as shown below.

If the farmer wants to pick one apple from each tree, how many apples will he pick?

_____

Lesson Check

1. The model shows how Maya planted flowers in her garden.

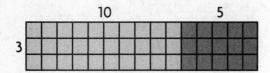

How many flowers did Maya plant?

Ⓐ 15

Ⓑ 18

Ⓒ 30

Ⓓ 45

2. The model below represents the expression 5×18.

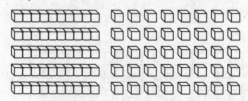

How many tens will there be in the final product?

Ⓐ 5   Ⓒ 8

Ⓑ 6   Ⓓ 9

Spiral Review

3. Center City has a population of twenty one thousand, seventy people. Which of the following shows the population written in standard form? **(Lesson 1.2)**

Ⓐ 21,007

Ⓑ 21,070

Ⓒ 21,077

Ⓓ 21,700

4. Central School collected 12,516 pounds of newspaper to recycle. Eastland School collected 12,615 pounds of newspapers. How many more pounds of newspaper did Eastland School collect than Central School? **(Lesson 1.7)**

Ⓐ   99 pounds

Ⓑ   101 pounds

Ⓒ   199 pounds

Ⓓ 1,099 pounds

5. Allison has 5 times as many baseball cards as football cards. In all, she has 120 baseball and football cards. How many baseball cards does Allison have?

(Lesson 2.2)

Ⓐ   20

Ⓑ   24

Ⓒ   96

Ⓓ 100

6. A ruby-throated hummingbird beats its wings about 53 times each second. About how many times does a ruby-throated hummingbird beat its wings in 5 seconds? **(Lesson 2.4)**

Ⓐ 25   Ⓒ 250

Ⓑ 58   Ⓓ 300

Name _____

Multiply Using Expanded Form

Record the product. Use expanded form to help.

1. $7 \times 14 =$ _____**98**_____

$7 \times 14 = 7 \times (10 + 4)$

$\qquad = (7 \times 10) + (7 \times 4)$

$\qquad = 70 + 28$

$\qquad = 98$

2. $8 \times 43 =$ _____

3. $6 \times 532 =$ _____

4. $5 \times 923 =$ _____

5. $4 \times 2,371 =$ _____

6. $7 \times 1,829 =$ _____

Problem Solving REAL WORLD

7. The fourth-grade students at Riverside School are going on a field trip. There are 68 students on each of the 4 buses. How many students are going on the field trip?

8. There are 5,280 feet in one mile. Hannah likes to walk 5 miles each week for exercise. How many feet does Hannah walk each week?

_____

_____

Lesson Check

1. Which expression shows how to multiply 7 × 256 by using expanded form and the Distributive Property?

   (A) (7 × 2) + (7 × 5) + (7 × 6)

   (B) (7 × 200) + (7 × 500) + (7 × 600)

   (C) (7 × 2) + (7 × 50) + (7 × 600)

   (D) (7 × 200) + (7 × 50) + (7 × 6)

2. Sue uses the expression (8 × 3,000) + (8 × 200) + (8 × 9) to help solve a multiplication problem. Which is Sue's multiplication problem?

   (A) 8 × 329

   (B) 8 × 3,029

   (C) 8 × 3,209

   (D) 8 × 3,290

Spiral Review

3. What is another way to write 9 × 200? (Lesson 1.5)

   (A) 18 ones

   (B) 18 tens

   (C) 18 hundreds

   (D) 18 thousands

4. What is the value of the digit 4 in 46,000? (Lesson 1.1)

   (A) 4 ten thousands

   (B) 4 thousands

   (C) 4 hundreds

   (D) 4 tens

5. Chris bought 6 packages of napkins for his restaurant. There were 200 napkins in each package. How many napkins did Chris buy? (Lesson 2.3)

   (A) 120

   (B) 1,200

   (C) 12,000

   (D) 120,000

6. Which of the following lists the numbers in order from **least** to **greatest**? (Lesson 1.3)

   (A) 8,512; 8,251; 8,125

   (B) 8,251; 8,125; 8,512

   (C) 8,125; 8,512; 8,251

   (D) 8,125; 8,251; 8,512

Name _____

Multiply Using Partial Products

Estimate. Then record the product.

1. Estimate: __1,200__

$$
\begin{array}{r}
243 \\
\times\ \ \ 6 \\
\hline
1,200 \\
240 \\
+\ \ \ \ 18 \\
\hline
1,458 \\
\end{array}
$$

2. Estimate: _____

$$
\begin{array}{r}
640 \\
\times\ \ \ 3 \\
\hline
\end{array}
$$

3. Estimate: _____

$$
\begin{array}{r}
\$149 \\
\times\ \ \ \ 5 \\
\hline
\end{array}
$$

4. Estimate: _____

$$
\begin{array}{r}
721 \\
\times\ \ \ 8 \\
\hline
\end{array}
$$

5. Estimate: _____

$$
\begin{array}{r}
293 \\
\times\ \ \ 4 \\
\hline
\end{array}
$$

6. Estimate: _____

$$
\begin{array}{r}
\$416 \\
\times\ \ \ \ 6 \\
\hline
\end{array}
$$

7. Estimate: _____

$$
\begin{array}{r}
961 \\
\times\ \ \ 2 \\
\hline
\end{array}
$$

8. Estimate: _____

$$
\begin{array}{r}
837 \\
\times\ \ \ 9 \\
\hline
\end{array}
$$

9. Estimate: _____

$$
\begin{array}{r}
652 \\
\times\ \ \ 4 \\
\hline
\end{array}
$$

10. Estimate: _____

$$
\begin{array}{r}
307 \\
\times\ \ \ 3 \\
\hline
\end{array}
$$

11. Estimate: _____

$$
\begin{array}{r}
543 \\
\times\ \ \ 7 \\
\hline
\end{array}
$$

12. Estimate: _____

$$
\begin{array}{r}
\$822 \\
\times\ \ \ \ 5 \\
\hline
\end{array}
$$

Problem Solving REAL WORLD

13. A maze at a county fair is made from 275 bales of hay. The maze at the state fair is made from 4 times as many bales of hay. How many bales of hay are used for the maze at the state fair?

_____

14. Pedro gets 8 hours of sleep each night. How many hours does Pedro sleep in a year with 365 days?

_____

Lesson Check

1. A passenger jet flies at an average speed of 548 miles per hour. At that speed, how many miles does the plane travel in 4 hours?

   (A) 2,092 miles

   (B) 2,112 miles

   (C) 2,192 miles

   (D) 2,480 miles

2. Use the model to find 3 × 157.

	100	50	7
3			

   (A) 300,171

   (B) 300,157

   (C) 471

   (D) 451

Spiral Review

3. The school fun fair made $1,768 on games and $978 on food sales. How much money did the fun fair make on games and food sales? (Lesson 1.6)

   (A) $2,636

   (B) $2,646

   (C) $2,736

   (D) $2,746

4. Use the table below.

State	Population
North Dakota	646,844
Alaska	698,473
Vermont	621,760

   Which of the following lists the states from least to greatest population? (Lesson 1.3)

   (A) Alaska, North Dakota, Vermont

   (B) Vermont, Alaska, North Dakota

   (C) North Dakota, Vermont, Alaska

   (D) Vermont, North Dakota, Alaska

5. A National Park covers 218,375 acres. What is this number written in expanded form? (Lesson 1.2)

   (A) 200,000 + 10,000 + 8,000 + 300 + 70 + 5

   (B) 20,000 + 1,000 + 800 + 30 + 75

   (C) 218 + 375

   (D) 218 thousand, 375

6. Last year a business had profits of $8,000. This year its profits are 5 times as great. What are this year's profits? (Lesson 2.3)

   (A) $4,000

   (B) $40,000

   (C) $44,000

   (D) $400,000

Name _____

Multiply Using Mental Math

Find the product. Tell which strategy you used.

1. 6×297 **Think:** $297 = 300 - 3$
$6 \times 297 = 6 \times (300 - 3)$
$= (6 \times 300) - (6 \times 3)$
$= 1,800 - 18$
$= 1,782$

1,782;

use subtraction

2. $8 \times 25 \times 23$ **3.** 8×604 **4.** 50×28

_____   _____   _____

_____   _____   _____

5. 9×199 **6.** $20 \times 72 \times 5$ **7.** 32×25

_____   _____   _____

_____   _____   _____

Problem Solving REAL WORLD

8. Section J in an arena has 20 rows. Each row has 15 seats. All tickets cost $18 each. If all the seats are sold, how much money will the arena collect for Section J?

9. At a high-school gym, the bleachers are divided into 6 equal sections. Each section can seat 395 people. How many people can be seated in the gym?

_____   _____

Lesson Check

1. Pencils come in cartons of 24 boxes. A school bought 50 cartons of pencils for the start of school. Each box of pencils cost $2. How much did the school spend on pencils?

   Ⓐ $240
   Ⓑ $1,200
   Ⓒ $2,400
   Ⓓ $4,800

2. The school also bought 195 packages of markers. There are 6 markers in a package. How many markers did the school buy?

   Ⓐ 1,170
   Ⓑ 1,195
   Ⓒ 1,200
   Ⓓ 1,230

Spiral Review

3. Alex has 175 baseball cards. Rodney has 3 times as many baseball cards as Alex. How many fewer cards does Alex have than Rodney? (Lesson 2.7)

   Ⓐ 700
   Ⓑ 525
   Ⓒ 450
   Ⓓ 350

4. A theater seats 1,860 people. The last 6 shows have been sold out. Which is the **best** estimate of the total number of people attending the last 6 shows?

   (Lesson 2.4)

   Ⓐ fewer than 6,000
   Ⓑ about 6,000
   Ⓒ fewer than 12,000
   Ⓓ more than 20,000

5. At one basketball game, there were 1,207 people watching. At the next game, there were 958 people. How many people in all were at the two games? (Lesson 1.6)

   Ⓐ 2,155
   Ⓑ 2,165
   Ⓒ 2,265
   Ⓓ 10,787

6. Bill bought 4 jigsaw puzzles. Each puzzle has 500 pieces. How many pieces are in all the puzzles altogether? (Lesson 2.3)

   Ⓐ 200
   Ⓑ 900
   Ⓒ 2,000
   Ⓓ 20,000

Name _____

Problem Solving • Multistep Multiplication Problems

Solve each problem.

1. A community park has 6 tables with a chessboard painted on top. Each board has 8 rows of 8 squares. When a game is set up, 4 rows of 8 squares on each board are covered with chess pieces. If a game is set up on each table, how many total squares are NOT covered by chess pieces?

   $4 \times 8 = 32$
   $32 \times 6 = \boxed{}$

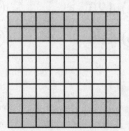

192 squares

2. Jonah and his friends go apple picking. Jonah fills 5 baskets. Each basket holds 15 apples. If 4 of Jonah's friends pick the same amount as Jonah, how many apples do Jonah and his friends pick in all? Draw a diagram to solve the problem.

   _____

3. There are 6 rows of 16 chairs set up for the third-grade play. In the first 4 rows, 2 chairs on each end are reserved for teachers. The rest of the chairs are for students. How many chairs are there for students?

   _____

Lesson Check

1. At a tree farm, there are 9 rows of 36 spruce trees. In each row, 14 of the spruce trees are blue spruce. How many spruce trees are NOT blue spruce?

   (A) 126   (C) 310

   (B) 198   (D) 324

2. Ron is tiling a countertop. He needs to place 54 square tiles in each of 8 rows to cover the counter. He wants to randomly place 8 groups of 4 blue tiles each and have the rest of the tiles be white. How many white tiles will Ron need?

   (A) 464   (C) 400

   (B) 432   (D) 32

Spiral Review

3. Juan reads a book with 368 pages. Savannah reads a book with 172 fewer pages than Juan's book. How many pages are in the book Savannah reads? (Lesson 1.8)

   (A) 196

   (B) 216

   (C) 296

   (D) 540

4. Hailey has bottles that hold 678 pennies each. About how many pennies does she have if she has 6 bottles filled with pennies? (Lesson 2.4)

   (A) 3,600

   (B) 3,900

   (C) 4,200

   (D) 6,000

5. Terrence plants a garden that has 8 rows of flowers, with 28 flowers in each row. How many flowers did Terrence plant? (Lesson 2.6)

   (A) 1,664

   (B) 224

   (C) 164

   (D) 36

6. Kevin has 5 fish in his fish tank. Jasmine has 4 times as many fish as Kevin has. How many fish does Jasmine have? (Lesson 2.1)

   (A) 15

   (B) 20

   (C) 25

   (D) 30

Name _____

Multiply 2-Digit Numbers with Regrouping

Estimate. Then record the product.

1. Estimate: __150__

   1
   46
   × 3
   ___
   138

2. Estimate: _____

   32
   × 8
   ___

3. Estimate: _____

   $55
   × 2
   ___

4. Estimate: _____

   61
   × 8
   ___

5. Estimate: _____

   37
   × 9
   ___

6. Estimate: _____

   $18
   × 7
   ___

7. Estimate: _____

   83
   × 5
   ___

8. Estimate: _____

   95
   × 8
   ___

9. Estimate: _____

   94
   × 9
   ___

10. Estimate: _____

    57
    × 6
    ___

11. Estimate: _____

    72
    × 3
    ___

12. Estimate: _____

    $79
    × 8
    ___

Problem Solving  REAL WORLD

13. Sharon is 54 inches tall. A tree in her backyard is 5 times as tall as she is. The floor of her treehouse is at a height that is twice as tall as she is. What is the difference, in inches, between the top of the tree and the floor of the treehouse?

14. Mr. Diaz's class is taking a field trip to the science museum. There are 23 students in the class, and a student admission ticket is $8. How much will the student tickets cost?

_____   _____

Lesson Check

1. A ferryboat makes four trips to an island each day. The ferry can hold 88 people. If the ferry is full on each trip, how many passengers are carried by the ferry each day?

   Ⓐ 176
   Ⓑ 322
   Ⓒ 332
   Ⓓ 352

2. Julian counted the number of times he drove across the Seven Mile Bridge while vacationing in the Florida Keys. He crossed the bridge 34 times. How many miles in all did Julian drive crossing the bridge?

   Ⓐ 328 miles   Ⓒ 238 miles
   Ⓑ 248 miles   Ⓓ 218 miles

Spiral Review

3. Sebastian wrote the population of his city as $300{,}000 + 40{,}000 + 60 + 7$. Which of the following shows the population of Sebastian's city written in standard form? (Lesson 1.2)

   Ⓐ 346,700
   Ⓑ 340,670
   Ⓒ 340,607
   Ⓓ 340,067

4. A plane flew 2,190 kilometers from Chicago to Flagstaff. Another plane flew 2,910 kilometers from Chicago to Oakland. How much farther did the plane that flew to Oakland fly than the plane that flew to Flagstaff? (Lesson 1.7)

   Ⓐ   720 kilometers
   Ⓑ   820 kilometers
   Ⓒ 5,000 kilometers
   Ⓓ 5,100 kilometers

5. Tori buys 27 packages of miniature racing cars. Each package contains 5 cars. About how many miniature racing cars does Tori buy? (Lesson 2.4)

   Ⓐ   15
   Ⓑ   32
   Ⓒ 100
   Ⓓ 150

6. Which of the following equations represents the Distributive Property? (Lesson 2.5)

   Ⓐ $3 \times 4 = 4 \times 3$
   Ⓑ $9 \times 0 = 0$
   Ⓒ $5 \times (3 + 4) = (5 \times 3) + (5 \times 4)$
   Ⓓ $6 \times (3 \times 2) = (6 \times 3) \times 2$

Multiply 3-Digit and 4-Digit Numbers with Regrouping

Name _____

Estimate. Then find the product.

1. Estimate: **4,000**

   1 2 2
$$\begin{array}{r} 1{,}467 \\ \times\ \ \ \ 4 \\ \hline 5{,}868 \end{array}$$

2. Estimate: _____

$$\begin{array}{r} 5{,}339 \\ \times\ \ \ \ 6 \\ \hline \end{array}$$

3. Estimate: _____

$$\begin{array}{r} \$879 \\ \times\ \ \ \ 8 \\ \hline \end{array}$$

4. Estimate: _____

$$\begin{array}{r} 3{,}182 \\ \times\ \ \ \ 5 \\ \hline \end{array}$$

5. Estimate: _____

$$\begin{array}{r} 4{,}616 \\ \times\ \ \ \ 3 \\ \hline \end{array}$$

6. Estimate: _____

$$\begin{array}{r} \$2{,}854 \\ \times\ \ \ \ 9 \\ \hline \end{array}$$

7. Estimate: _____

$$\begin{array}{r} 7{,}500 \\ \times\ \ \ \ 2 \\ \hline \end{array}$$

8. Estimate: _____

$$\begin{array}{r} 948 \\ \times\ \ \ \ 7 \\ \hline \end{array}$$

9. Estimate: _____

$$\begin{array}{r} 1{,}752 \\ \times\ \ \ \ 6 \\ \hline \end{array}$$

10. Estimate: _____

$$\begin{array}{r} 550 \\ \times\ \ \ \ 9 \\ \hline \end{array}$$

11. Estimate: _____

$$\begin{array}{r} 6{,}839 \\ \times\ \ \ \ 4 \\ \hline \end{array}$$

12. Estimate: _____

$$\begin{array}{r} \$9{,}614 \\ \times\ \ \ \ 3 \\ \hline \end{array}$$

Problem Solving  REAL WORLD

13. Lafayette County has a population of 7,022 people. Columbia County's population is 8 times as great as Lafayette County's population. What is the population of Columbia County?

14. A seafood company sold 9,125 pounds of fish last month. If 6 seafood companies sold the same amount of fish, how much fish did the 6 companies sell last month in all?

_____

_____

Lesson Check

1. By recycling 1 ton of paper, 6,953 gallons of water are saved. How many gallons of water are saved by recycling 4 tons of paper?

   (A) 24,602 gallons

   (B) 27,612 gallons

   (C) 27,812 gallons

   (D) 28,000 gallons

2. Esteban counted the number of steps it took him to walk to school. He counted 1,138 steps. How many steps does he take walking to and from school each day?

   (A) 2,000

   (B) 2,266

   (C) 2,276

   (D) 22,616

Spiral Review

3. A website has 13,406 people registered. What is the word form of this number? (Lesson 1.2)

   (A) thirty thousand, four hundred six

   (B) thirteen thousand, four hundred sixty

   (C) thirteen thousand, four hundred six

   (D) thirteen thousand, six hundred six

4. In one year, the McAlister family drove their car 15,680 miles. To the nearest thousand, how many miles did they drive their car that year? (Lesson 1.4)

   (A) 15,000 miles

   (B) 15,700 miles

   (C) 16,000 miles

   (D) 20,000 miles

5. Connor scored 14,370 points in a game. Amy scored 1,089 fewer points than Connor. How many points did Amy score? (Lesson 1.8)

   (A) 12,281

   (B) 13,281

   (C) 15,359

   (D) 15,459

6. Lea buys 6 model cars that each cost $15. She also buys 4 bottles of paint that each cost $11. How much does Lea spend in all on model cars and paint? (Lesson 2.9)

   (A) $134

   (B) $90

   (C) $44

   (D) $36

Solve Multistep Problems Using Equations

Find the value of *n*.

1. $4 \times 27 + 5 \times 34 - 94 = n$

   $108 + 5 \times 34 - 94 = n$

   $108 + 170 - 94 = n$

   $278 - 94 = n$

   $184 = n$

2. $7 \times 38 + 3 \times 45 - 56 = n$

   _____ $= n$

3. $6 \times 21 + 7 \times 29 - 83 = n$

   _____ $= n$

4. $9 \times 19 + 2 \times 57 - 75 = n$

   _____ $= n$

5. $5 \times 62 + 6 \times 33 - 68 = n$

   _____ $= n$

6. $8 \times 19 + 4 \times 49 - 39 = n$

   _____ $= n$

Problem Solving REAL WORLD

7. A bakery has 4 trays with 16 muffins on each tray. The bakery has 3 trays of cupcakes with 24 cupcakes on each tray. If 15 cupcakes are sold, how many muffins and cupcakes are left?

   _____

   _____

8. Katy bought 5 packages of stickers with 25 stickers in each package. She also bought 3 boxes of markers with 12 markers in each box. If she receives 8 stickers from a friend, how many stickers and markers does Katy have now?

   _____

   _____

Lesson Check

1. What is the value of n?

   $9 \times 23 + 3 \times 39 - 28 = n$

   (A) 240

   (B) 296

   (C) 2,310

   (D) 8,162

2. Which expression has a value of 199?

   (A) $4 \times 28 + 6 \times 17 - 15$

   (B) $4 \times 17 + 6 \times 28 - 38$

   (C) $4 \times 38 + 6 \times 15 - 28$

   (D) $4 \times 15 + 6 \times 38 - 88$

Spiral Review

3. Which expression shows how you can multiply 9×475 using expanded form and the Distributive Property? (Lesson 2.6)

   (A) $(9 \times 4) + (9 \times 7) + (9 \times 5)$

   (B) $(9 \times 4) + (9 \times 70) + (9 \times 700)$

   (C) $(9 \times 400) + (9 \times 70) + (9 \times 5)$

   (D) $(9 \times 400) + (9 \times 700) + (9 \times 500)$

4. Which equation best represents the comparison sentence? (Lesson 2.1)

   32 is 8 times as many as 4

   (A) $32 = 8 \times 4$

   (B) $32 \times 8 = 4$

   (C) $32 = 8 + 4$

   (D) $8 + 4 = 32$

5. Between which pair of numbers is the exact product of 379 and 8? (Lesson 2.4)

   (A) between 2,400 and 2,500

   (B) between 2,400 and 2,800

   (C) between 2,400 and 3,000

   (D) between 2,400 and 3,200

6. Which of the following statements shows the halving and doubling strategy to find 28×50? (Lesson 2.8)

   (A) $28 \times 50 = 14 \times 100$

   (B) $28 \times 50 = (14 \times 25) \times (14 \times 25)$

   (C) $28 \times 50 = (20 \times 50) + (8 \times 50)$

   (D) $28 \times 50 = 2 \times (14 \times 25)$

Name _____

Chapter 2 Extra Practice

Lesson 2.1

Write a comparison sentence.

1. $27 = 3 \times 9$

   _____ is _____ times as many as _____.

2. $7 \times 8 = 56$

   _____ times as many as _____ is _____.

Lessons 2.3, 2.5 - 2.6

Find the product.

1. $2 \times 700 =$ _____

2. $6 \times 6,000 =$ _____

3. $7 \times 13 =$ _____

4. $4 \times 19 =$ _____

5. $5 \times 216 =$ _____

6. $9 \times 1,362 =$ _____

Lessons 2.2, 2.9

Draw a diagram. Write an equation and solve.

1. Julia saw 5 times as many cars as trucks in a parking lot. If she saw 30 cars and trucks altogether in the parking lot, how many were trucks?

   _____

   _____

2. Ivan has 6 times as many blue beads as red beads. He has 49 red and blue beads in all. How many blue beads does Ivan have?

   _____

   _____

3. There are 6 rows with 18 chairs in each row. In the center of the chairs, 4 rows of 6 chairs are brown. The rest of the chairs are blue. How many chairs are blue?

   _____

   _____

Lessons 2.7, 2.10 - 2.11

Estimate. Then record the product.

1. Estimate: _____

$$\begin{array}{r} 318 \\ \times\ \ 3 \\ \hline \end{array}$$

2. Estimate: _____

$$\begin{array}{r} \$522 \\ \times\ \ 9 \\ \hline \end{array}$$

3. Estimate: _____

$$\begin{array}{r} \$36 \\ \times\ 6 \\ \hline \end{array}$$

4. Estimate: _____

$$\begin{array}{r} 57 \\ \times\ 8 \\ \hline \end{array}$$

5. Estimate: _____

$$\begin{array}{r} 3,600 \\ \times\ \ \ \ 8 \\ \hline \end{array}$$

6. Estimate: _____

$$\begin{array}{r} \$9,107 \\ \times\ \ \ \ 5 \\ \hline \end{array}$$

Lesson 2.8

Find the product. Tell which strategy you used.

1. $(4 \times 10) \times 10$

2. 2×898

3. $4 \times 7 \times 25$

_____   _____   _____

_____   _____   _____

Lessons 2.4, 2.12

1. School pennants cost $18 each. Ms. Lee says she will pay $146 for 7 pennants. Is her answer reasonable? Explain.

2. Caleb draws 14 dogs on each of 4 posters. He draws 18 cats on each of 6 other posters. If he draws 5 more dogs on each poster with dogs, how many dogs and cats does he draw?

_____   _____

School-Home Letter

Dear Family,

During the next few weeks, our math class will be learning to multiply by 2-digit whole numbers. We will also learn how to describe the reasonableness of an estimate.

You can expect to see homework that provides practice with estimation and multiplication of numbers with more than 1 digit.

Here is a sample of how your child will be taught to multiply by a 2-digit number.

Vocabulary

compatible numbers Numbers that are easy to compute mentally

estimate To find an answer that is close to the exact amount

partial products A method of multiplying in which the ones, tens, hundreds, and so on are multiplied separately and then the products are added together

---

🔒 MODEL Multiply 2-Digit Numbers

This is one way that we will be multiplying by 2-digit numbers.

STEP 1

Multiply by the ones digit.

$$
\begin{array}{r}
\overset{2}{2}4 \\
\times\ 25 \\
\hline
120
\end{array}
\leftarrow
\begin{array}{l}
\text{partial} \\
\text{product}
\end{array}
$$

STEP 2

Multiply by the tens digit. Start by placing a zero in the ones place.

$$
\begin{array}{r}
\overset{2}{2}4 \\
\times\ 25 \\
\hline
120 \\
+\ 480
\end{array}
\leftarrow
\begin{array}{l}
\text{partial} \\
\text{product}
\end{array}
$$

STEP 3

Add the partial products.

$$
\begin{array}{r}
\overset{2}{2}4 \\
\times\ 25 \\
\hline
120 \\
+\ 480 \\
\hline
600
\end{array}
\leftarrow \text{product}
$$

Tips

Estimating to Check Multiplication

When estimation is used to check that a multiplication answer is reasonable, usually each factor is rounded to a multiple of 10 that has only one nonzero digit. Then mental math can be used to recall the basic fact product, and patterns can be used to determine the correct number of zeros in the estimate.

Capítulo 3 — Carta para la casa

Querida familia,

Durante las próximas semanas, en la clase de matemáticas aprenderemos a multiplicar por números enteros de 2 dígitos. También aprenderemos cómo describir qué tan razonable es una estimación.

Llevaré a la casa tareas con actividades para practicar la estimación y la multiplicación de números con más de 1 dígito.

Este es un ejemplo de la manera como aprenderemos a multiplicar por números de 2 dígitos.

Vocabulario

números compatibles Números que son fáciles de calcular mentalmente

estimar Hallar un total que se aproxime a la cantidad exacta

productos parciales Método de multiplicación a través del cual las unidades, decenas, centenas, etc., se multiplican por separado, y luego se suman los productos

🔑 MODELO Multiplicar números de 2 dígitos

Esta es una manera en la que multiplicaremos por números de 2 dígitos.

PASO 1

Multiplica por el dígito de las unidades.

$$
\begin{array}{r}
\overset{2}{2}4 \\
\times\ 25 \\
\hline
120
\end{array}
$$
← producto parcial

PASO 2

Multiplica por el dígito de las decenas. Comienza escribiendo un cero en el lugar del las unidades.

$$
\begin{array}{r}
\overset{2}{2}4 \\
\times\ 25 \\
\hline
120 \\
+\ 480
\end{array}
$$
← producto parcial

PASO 3

Suma los productos parciales.

$$
\begin{array}{r}
\overset{2}{2}4 \\
\times\ 25 \\
\hline
120 \\
+\ 480 \\
\hline
600
\end{array}
$$
← producto

Pistas

Estimar para comprobar la multiplicación

Cuando se usa la estimación para comprobar que la respuesta de una multiplicación es razonable, cada factor se suele redondear al múltiplo de 10 que tiene un solo dígito distinto de cero. Después se puede usar el cálculo mental para recordar el producto básico de la operación, y se pueden usar patrones para determinar la cantidad correcta de ceros de la estimación.

© Houghton Mifflin Harcourt Publishing Company

Name _____

Multiply by Tens

Choose a method. Then find the product.

1. 16×60

Use the halving-and-doubling strategy.

Find half of 16: $16 \div 2 = 8$.

Multiply this number by 60: $8 \times 60 = 480$

Double this result: $2 \times 480 = 960$

__960__

2. 80×22

3. 30×52

4. 60×20

_____

_____

_____

5. 40×35

6. 10×90

7. 31×50

_____

_____

_____

Problem Solving REAL WORLD

8. Kenny bought 20 packs of baseball cards. There are 12 cards in each pack. How many cards did Kenny buy?

9. The Hart family drove 10 hours to their vacation spot. They drove an average of 48 miles each hour. How many miles did they drive in all?

_____

_____

Lesson Check

1. For the school play, 40 rows of chairs are set up. There are 22 chairs in each row. How many chairs are there in all?

   (A) 800

   (B) 840

   (C) 880

   (D) 8,800

2. At West School, there are 20 classrooms. Each classroom has 20 students. How many students are at West School?

   (A) 40

   (B) 400

   (C) 440

   (D) 4,000

Spiral Review

3. Alex has 48 stickers. This is 6 times the number of stickers Max has. How many stickers does Max have? (Lesson 2.1)

   (A) 6

   (B) 7

   (C) 8

   (D) 9

4. Ali's dog weighs 8 times as much as her cat. Together, the two pets weigh 54 pounds. How much does Ali's dog weigh? (Lesson 2.2)

   (A) 6 pounds

   (B) 42 pounds

   (C) 46 pounds

   (D) 48 pounds

5. Allison has 3 containers with 25 crayons in each. She also has 4 boxes of markers with 12 markers in each box. She gives 10 crayons to a friend. How many crayons and markers does Allison have now?

   (Lesson 2.12)

   (A) 34

   (B) 113

   (C) 123

   (D) 133

6. The state of Utah covers 82,144 square miles. The state of Montana covers 145,552 square miles. What is the total area of the two states? (Lesson 1.6)

   (A) 63,408 square miles

   (B) 223,408 square miles

   (C) 227,696 square miles

   (D) 966,992 square miles

Name _____

Estimate Products

Estimate the product. Choose a method.

1. 38 × 21

38 × 21

40 × 20

800

_____

2. 63 × 19

_____

3. 27 × $42

_____

4. 73 × 67

_____

5. 37 × $44

_____

6. 85 × 71

_____

7. 88 × 56

_____

8. 97 × 13

_____

9. 92 × 64

_____

Problem Solving

10. A dime has a diameter of about 18 millimeters. About how many millimeters long would a row of 34 dimes be?

11. A half-dollar has a diameter of about 31 millimeters. About how many millimeters long would a row of 56 half-dollars be?

_____

_____

Lesson Check

1. Which is the best estimate for the product 43 × 68?

    (A) 3,500

    (B) 2,800

    (C) 2,400

    (D) 280

2. Marissa burns 93 calories each time she plays fetch with her dog. She plays fetch with her dog once a day. About how many calories will Marissa burn playing fetch with her dog in 28 days?

    (A) 4,000   (C) 2,000

    (B) 2,700   (D) 270

Spiral Review

3. Use the model to find 3 × 126. (Lesson 2.7)

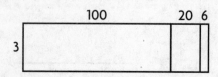

    (A) 368

    (B) 378

    (C) 468

    (D) 478

4. A store sells a certain brand of jeans for $38. One day, the store sold 6 pairs of jeans of that brand. How much money did the store make from selling the 6 pairs of jeans? (Lesson 2.10)

    (A) $188

    (B) $228

    (C) $248

    (D) $288

5. The Gateway Arch in St. Louis, Missouri, weighs about 20,000 tons. Which amount could be the exact number of tons the Arch weighs? (Lesson 1.4)

    (A) 31,093 tons

    (B) 25,812 tons

    (C) 17,246 tons

    (D) 14,096 tons

6. Which is another name for 23 ten thousands? (Lesson 1.5)

    (A) 23,000,000

    (B) 2,300,000

    (C) 230,000

    (D) 23,000

Name _____

Area Models and Partial Products

Draw a model to represent the product.
Then record the product.

1. 13 × 42

	40	2
10	400	20
3	120	6

400 + 20 + 120 + 6 = **546** _____

2. 18 × 34

_____

3. 22 × 26

_____

4. 15 × 33

_____

5. 23 × 29

_____

6. 19 × 36

_____

Problem Solving  REAL WORLD

7. Sebastian made the following model to find the product 17 × 24.

	20	4
10	200	40
7	14	28

200 + 40 + 14 + 28 = 282

Is his model correct? **Explain.**

_____

_____

8. Each student in Ms. Sike's kindergarten class has a box of crayons. Each box has 36 crayons. If there are 18 students in Ms. Sike's class, how many crayons are there in all?

_____

_____

Lesson Check

1. Which product does the model below represent?

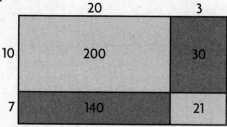

(A) 161   (C) 340

(B) 230   (D) 391

2. Which product does the model below represent?

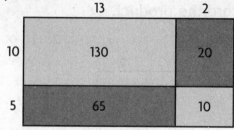

(A) 219   (C) 244

(B) 225   (D) 275

Spiral Review

3. Mariah builds a tabletop using square tiles. There are 12 rows of tiles and 30 tiles in each row. How many tiles in all does Mariah use? (Lesson 3.1)

(A) 100

(B) 180

(C) 360

(D) 420

4. Trevor bakes 8 batches of biscuits, with 14 biscuits in each batch. He sets aside 4 biscuits from each batch for a bake sale and puts the rest in a jar. How many biscuits does Trevor put in the jar?

(Lesson 2.12)

(A) 112

(B) 80

(C) 50

(D) 32

5. Li feeds her dog 3 cups of food each day. About how many cups of food does her dog eat in 28 days? (Lesson 2.4)

(A) 60 cups   (C) 80 cups

(B) 70 cups   (D) 90 cups

6. Which symbol makes the number sentence true? (Lesson 2.8)

$$4 \boxed{} 0 = 0$$

(A) +   (C) ×

(B) −   (D) ÷

Name _____

Multiply Using Partial Products

Record the product.

1.   23
  × 79

1,400
210
180
+ 27
1,817

2.   56
  × 32

3.   87
  × 64

4.   33
  × 25

5.   94
  × 12

6.   51
  × 77

7.   69
  × 49

8.   86
  × 84

9.   98
  × 42

10.   73
  × 37

11.   85
  × 51

Problem Solving  REAL WORLD

12. Evelyn drinks 8 glasses of water a day, which is 56 glasses of water a week. How many glasses of water does she drink in a year? (1 year = 52 weeks)

13. Joe wants to use the Hiking Club's funds to purchase new walking sticks for each of its 19 members. The sticks cost $26 each. The club has $480. Is this enough money to buy each member a new walking stick? If not, how much more money is needed?

Lesson Check

1. A carnival snack booth made $76 selling popcorn in one day. It made 22 times as much selling cotton candy. How much money did the snack booth make selling cotton candy?

   (A) $284   (C) $1,562

   (B) $304   (D) $1,672

2. What are the partial products of 42 × 28?

   (A) 800, 80, 40, 16

   (B) 800, 16

   (C) 800, 40, 320, 16

   (D) 80, 16

Spiral Review

3. Last year, the city library collected 117 used books for its shelves. This year, it collected 3 times as many books. How many books did it collect this year?
   (Lesson 2.1)

   (A) 832

   (B) 428

   (C) 351

   (D) 72

4. Washington Elementary has 232 students. Washington High has 6 times as many students. How many students does Washington High have?
   (Lesson 2.11)

   (A) 1,392

   (B) 1,382

   (C) 1,292

   (D) 1,281

5. What are the partial products of 35 × 7?
   (Lesson 2.7)

   (A) 10, 12

   (B) 21, 35

   (C) 210, 35

   (D) 350, 21

6. Shelby has ten $5 bills and thirteen $10 bills. How much money does Shelby have in all? (Lesson 2.12)

   (A) $15

   (B) $60

   (C) $63

   (D) $180

Name _____

Multiply with Regrouping

Estimate. Then find the product.

1. Estimate: ___2,700___

$$\begin{array}{r} \overset{2}{\overset{1}{}} \\ 87 \\ \times\ \ 32 \\ \hline 174 \\ +\ 2,610 \\ \hline 2,784 \end{array}$$

Think: 87 is close to 90 and 32 is close to 30.

$$90 \times 30 = 2,700$$

2. Estimate: _____

$$\begin{array}{r} 73 \\ \times\ \ 28 \\ \hline \end{array}$$

3. Estimate: _____

$$\begin{array}{r} 48 \\ \times\ \ 38 \\ \hline \end{array}$$

4. Estimate: _____

$$\begin{array}{r} 59 \\ \times\ \ 52 \\ \hline \end{array}$$

5. Estimate: _____

$$\begin{array}{r} 84 \\ \times\ \ 40 \\ \hline \end{array}$$

6. Estimate: _____

$$\begin{array}{r} 83 \\ \times\ \ 77 \\ \hline \end{array}$$

7. Estimate: _____

$$\begin{array}{r} 91 \\ \times\ \ 19 \\ \hline \end{array}$$

Problem Solving REAL WORLD

8. Baseballs come in cartons of 84 baseballs. A team orders 18 cartons of baseballs. How many baseballs does the team order?

9. There are 16 tables in the school lunch room. Each table can seat 22 students. How many students can be seated at lunch at one time?

_____

_____

Lesson Check

1. The art teacher has 48 boxes of crayons. There are 64 crayons in each box. Which is the best estimate of the number of crayons the art teacher has?

   (A) 2,400

   (B) 2,800

   (C) 3,000

   (D) 3,500

2. A basketball team scored an average of 52 points in each of 15 games. How many points did the team score in all?

   (A) 500

   (B) 312

   (C) 780

   (D) 1,000

Spiral Review

3. One Saturday, an orchard sold 83 bags of apples. There are 27 apples in each bag. Which expression represents the total number of apples sold? (Lesson 3.4)

   (A) 16 + 6 + 56 + 21

   (B) 160 + 60 + 56 + 21

   (C) 160 + 60 + 560 + 21

   (D) 1,600 + 60 + 560 + 21

4. Hannah has a grid of squares that has 12 rows with 15 squares in each row. She colors 5 rows of 8 squares in the middle of the grid blue. She colors the rest of the squares red. How many squares does Hannah color red? (Lesson 2.9)

   (A) 40

   (B) 140

   (C) 180

   (D) 220

5. Gabriella has 4 times as many erasers a Leona. Leona has 8 erasers. How many erasers does Gabriella have? (Lesson 2.1)

   (A) 32

   (B) 24

   (C) 12

   (D) 2

6. Phil has 3 times as many rocks as Peter. Together, they have 48 rocks. How many more rocks does Phil have than Peter?

   (Lesson 2.2)

   (A) 36

   (B) 24

   (C) 16

   (D) 12

Choose a Multiplication Method

Estimate. Then choose a method to find the product.

1. Estimate: **1,200**

$$
\begin{array}{r}
31 \\
\times\ 43 \\
\hline
93 \\
+\ 1{,}240 \\
\hline
1{,}333
\end{array}
$$

2. Estimate: _____

$$
\begin{array}{r}
67 \\
\times\ 85 \\
\hline
\end{array}
$$

3. Estimate: _____

$$
\begin{array}{r}
68 \\
\times\ 38 \\
\hline
\end{array}
$$

4. Estimate: _____

$$
\begin{array}{r}
95 \\
\times\ 17 \\
\hline
\end{array}
$$

5. Estimate: _____

$$
\begin{array}{r}
49 \\
\times\ 54 \\
\hline
\end{array}
$$

6. Estimate: _____

$$
\begin{array}{r}
91 \\
\times\ 26 \\
\hline
\end{array}
$$

7. Estimate: _____

$$
\begin{array}{r}
82 \\
\times\ 19 \\
\hline
\end{array}
$$

8. Estimate: _____

$$
\begin{array}{r}
46 \\
\times\ 27 \\
\hline
\end{array}
$$

9. Estimate: _____

$$
\begin{array}{r}
41 \\
\times\ 33 \\
\hline
\end{array}
$$

10. Estimate: _____

$$
\begin{array}{r}
97 \\
\times\ 13 \\
\hline
\end{array}
$$

11. Estimate: _____

$$
\begin{array}{r}
75 \\
\times\ 69 \\
\hline
\end{array}
$$

Problem Solving REAL WORLD

12. A movie theatre has 26 rows of seats. There are 18 seats in each row. How many seats are there in all?

_____

13. Each class at Briarwood Elementary collected at least 54 cans of food during the food drive. If there are 29 classes in the school, what was the least number of cans collected?

_____

Lesson Check

1. A choir needs new robes for each of its 46 singers. Each robe costs $32. What will be the total cost for all 46 robes?

- (A) $1,472
- (B) $1,372
- (C) $1,362
- (D) $230

2. A wall on the side of a building is made up of 52 rows of bricks with 44 bricks in each row. How many bricks make up the wall?

- (A) 3,080
- (B) 2,288
- (C) 488
- (D) 416

Spiral Review

3. Which expression shows how to multiply 4×362 by using place value and expanded form? (Lesson 2.6)

- (A) $(4 \times 3) + (4 \times 6) + (4 \times 2)$
- (B) $(4 \times 300) + (4 \times 600) + (4 \times 200)$
- (C) $(4 \times 300) + (4 \times 60) + (4 \times 20)$
- (D) $(4 \times 300) + (4 \times 60) + (4 \times 2)$

4. Use the model below. What is the product 4×492? (Lesson 2.7)

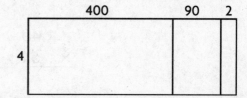

- (A) $16 + 36 + 8 = 60$
- (B) $160 + 36 + 8 = 204$
- (C) $160 + 360 + 8 = 528$
- (D) $1,600 + 360 + 8 = 1,968$

5. What is the sum $13,094 + 259,728$?
(Lesson 1.6)

- (A) 272,832
- (B) 272,822
- (C) 262,722
- (D) 262,712

6. During the 2008–2009 season, there were 801,372 people who attended the home hockey games in Philadelphia. There were 609,907 people who attended the home hockey games in Phoenix. How much greater was the home attendance in Philadelphia than in Phoenix that season? (Lesson 1.7)

- (A) 101,475
- (B) 191,465
- (C) 201,465
- (D) 202,465

Name _____

Problem Solving • Multiply 2-Digit Numbers

Solve each problem. Use a bar model to help.

1. Mason counted an average of 18 birds at his bird feeder each day for 20 days. Gloria counted an average of 21 birds at her bird feeder each day for 16 days. How many more birds did Mason count at his feeder than Gloria counted at hers?

360 birds counted by Mason

336 birds counted by Gloria

?

   Birds counted by Mason: 18 × 20 = 360

   Birds counted by Gloria: 21 × 16 = 336

   Draw a bar model to compare.

   Subtract. 360 − 336 = 24

   So, Mason counted __**24**__ more birds.

2. The 24 students in Ms. Lee's class each collected an average of 18 cans for recycling. The 21 students in Mr. Galvez's class each collected an average of 25 cans for recycling. How many more cans were collected by Mr. Galvez's class than Ms. Lee's class?

   _____

3. At East School, each of the 45 classrooms has an average of 22 students. At West School, each of the 42 classrooms has an average of 23 students. How many more students are at East School than at West School?

   _____

4. A zoo gift shop orders 18 boxes of 75 key rings each and 15 boxes of 80 refrigerator magnets each. How many more key rings than refrigerator magnets does the gift shop order?

   _____

Lesson Check

1. Ace Manufacturing ordered 17 boxes with 85 ball bearings each. They also ordered 15 boxes with 90 springs each. How many more ball bearings than springs did they order?

   (A) 5

   (B) 85

   (C) 90

   (D) 95

2. Elton hiked 16 miles each day on a 12-day hiking trip. Lola hiked 14 miles each day on her 16-day hiking trip. In all, how many more miles did Lola hike than Elton hiked?

   (A)   2 miles

   (B) 18 miles

   (C) 32 miles

   (D) 118 miles

Spiral Review

3. An orchard has 24 rows of apple trees. There are 35 apple trees in each row. How many apple trees are in the orchard? (Lesson 3.6)

   (A) 59

   (B) 192

   (C) 740

   (D) 840

4. An amusement park reported 354,605 visitors last summer. What is this number rounded to the nearest thousand? (Lesson 1.4)

   (A) 354,600

   (B) 355,000

   (C) 360,000

   (D) 400,000

5. Attendance at the football game was 102,653. What is the value of the digit 6?
   (Lesson 1.1)

   (A)   6

   (B)   60

   (C)   600

   (D) 6,000

6. Jill's fish weighs 8 times as much as her parakeet. Together, the pets weigh 63 ounces. How much does the fish weigh? (Lesson 2.2)

   (A) 7 ounces

   (B) 49 ounces

   (C) 55 ounces

   (D) 56 ounces

Name _____

Chapter 3 Extra Practice

Lesson 3.1

Choose a method. Then find the product.

1. 12×60 **2.** 56×40 **3.** 30×40 **4.** 50×67

_____ _____ _____ _____

Lesson 3.2

Estimate the product. Choose a method.

1. 33×76 **2.** 43×90 **3.** $47 \times \$66$ **4.** 12×81

_____ _____ _____ _____

5. 46×47 **6.** 58×79 **7.** 24×73 **8.** 68×36

_____ _____ _____ _____

Lesson 3.3

Draw a model to represent the product.
Then record the product.

1. 41×16 **2.** 39×52 **3.** 94×36

_____ _____ _____

Lesson 3.4

Record the product.

1.	2.	3.	4.
53 × 37	48 × 47	65 × 28	92 × 79

Lessons 3.5 – 3.6

Estimate. Then choose a method to find the product.

1. Estimate: _____

$$\begin{array}{r} 48 \\ \times\ 21 \\ \hline \end{array}$$

2. Estimate: _____

$$\begin{array}{r} \$\ 72 \\ \times\ \ \ 46 \\ \hline \end{array}$$

3. Estimate: _____

$$\begin{array}{r} 39 \\ \times\ 58 \\ \hline \end{array}$$

4. $27 \times \$19$

5. 97×32

6. 44×69

_____   _____   _____

Lesson 3.7

1. Last week, Ms. Simpson worked 28 hours. She stocked shelves for 45 minutes each hour for 14 of those hours. The rest of the time she worked in customer service. How many minutes last week did Ms. Simpson work in customer service?
   (Hint: 1 hour = 60 minutes)

2. The after-school craft center has 15 boxes of 64 crayons each. In 12 of the boxes, 28 of the crayons have not been used. All the rest have been used. How many of the crayons in the center have been used?

_____   _____

School-Home Letter

© Houghton Mifflin Harcourt Publishing Company

Dear Family,

During the next few weeks, our math class will be learning how to model division, and use the division algorithm to divide up to three-digit dividends by 1-digit divisors. The class will learn different methods to divide, including using models, repeated subtraction, and the standard division algorithm. We will also learn to divide with remainders.

You can expect to see homework that provides practice modeling division and using the division algorithm.

Here is a sample of how your child will be taught to model division using the Distributive Property.

Vocabulary

Distributive Property The property that states that dividing a sum by a number is the same as dividing each addend by the number and then adding the quotients

multiple A number that is the product of a given number and a counting number

remainder The amount left over when a number cannot be divided evenly

🔒 MODEL Use the Distributive Property to Divide

This is how we will divide using the Distributive Property.

Find $72 \div 3$.

STEP 1

Draw a rectangle to model $72 \div 3$.

```
        ?
   ┌──────────┐
 3 │    72    │
   └──────────┘
```

STEP 2

Think of 72 as $60 + 12$. Break apart the model into two rectangles to show $(60 + 12) \div 3$.

```
   ┌──────┬─────┐
 3 │  60  │ 12  │
   └──────┴─────┘
```

Tips

Whenever possible, try to use division facts and multiples of ten when breaking your rectangle into smaller rectangles. In the problem at the left, $60 \div 3$ is easy to find mentally.

STEP 3

Each rectangle models a division.

$$72 \div 3 = (60 \div 3) + (12 \div 3)$$
$$= 20 + 4$$
$$= 24$$

So, $72 \div 3 = 24$.

Carta
para la casa

Vocabulario

propiedad distributiva La propiedad que establece que dividir una suma entre un número es lo mismo que dividir cada sumando entre el número y luego sumar los cocientes

múltiplo Un número que es el producto de un número determinado y de un número positivo distinto de cero

residuo La cantidad sobrante cuando un número no se puede dividir en partes iguales

Querida familia,

Durante las próximas semanas, en la clase de matemáticas aprenderemos a representar la división y a usar el algoritmo de la división para dividir dividendos de hasta tres dígitos entre divisores de un dígito. Para ello, desarrollaremos diferentes métodos para dividir, incluyendo usar modelos, resta repetida y el algoritmo de la división estándar. También aprenderemos a dividir con residuos.

Llevaré a la casa tareas con actividades para representar la división y para usar el algoritmo de la división.

Este es un ejemplo de la manera como aprenderemos a representar la división usando la propiedad distributiva.

 MODELO Usar la propiedad distributiva para dividir

Así es como dividiremos usando la propiedad distributiva.

Halla $72 \div 3$.

PASO 1

Dibuja un rectángulo para representar $72 \div 3$.

?

3 | 72

PASO 2

Piensa en 72 como $60 + 12$.
Divide el modelo en dos rectángulos para mostrar $(60 + 12) \div 3$.

3 | 60 | 12

> **Pistas**
>
> En la medida de lo posible, trata de usar operaciones de división y múltiplos de diez cuando dividas el modelo en rectángulos más pequeños. En el problema anterior, $60 \div 3$ es fácil de hallar mentalmente.

PASO 3

Cada rectángulo representa una división.

$$72 \div 3 = (60 \div 3) + (12 \div 3)$$
$$= 20 + 4$$
$$= 24$$

Por tanto, $72 \div 3 = 24$.

Name _____

Estimate Quotients Using Multiples

Find two numbers the quotient is between. Then estimate the quotient.

1. $175 \div 6$

<u>between 20 and 30</u>

<u>about 30</u>

Think: $6 \times 20 = 120$ and $6 \times 30 = 180$.
So, $175 \div 6$ is between 20 and 30. Since 175 is closer to 180 than to 120, the quotient is about 30.

2. $53 \div 3$

_____

_____

3. $75 \div 4$

_____

_____

4. $215 \div 9$

_____

_____

5. $284 \div 5$

_____

_____

6. $191 \div 3$

_____

_____

7. $100 \div 7$

_____

_____

8. $438 \div 7$

_____

_____

9. $103 \div 8$

_____

_____

10. $255 \div 9$

_____

_____

Problem Solving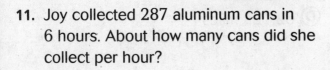

11. Joy collected 287 aluminum cans in 6 hours. About how many cans did she collect per hour?

_____

12. Paul sold 162 cups of lemonade in 5 hours. About how many cups of lemonade did he sell each hour?

_____

Lesson Check

1. Abby did 121 sit-ups in 8 minutes. Which is the best estimate of the number of sit-ups she did in 1 minute?

   (A) about 12

   (B) about 15

   (C) about 16

   (D) about 20

2. The Garibaldi family drove 400 miles in 7 hours. Which is the best estimate of the number of miles they drove in 1 hour?

   (A) about 40 miles

   (B) about 50 miles

   (C) about 60 miles

   (D) about 70 miles

Spiral Review

3. Twelve boys collected 16 aluminum cans each. Fifteen girls collected 14 aluminum cans each. How many more cans did the girls collect than the boys? (Lesson 3.7)

   (A) 8

   (B) 12

   (C) 14

   (D) 18

4. George bought 30 packs of football cards. There were 14 cards in each pack. How many cards did George buy? (Lesson 3.1)

   (A) 170

   (B) 320

   (C) 420

   (D) 520

5. Sarah made a necklace using 5 times as many blue beads as white beads. She used a total of 30 beads. How many blue beads did Sarah use? (Lesson 2.2)

   (A) 5

   (B) 6

   (C) 24

   (D) 25

6. This year, Ms. Webster flew 145,000 miles on business. Last year, she flew 83,125 miles on business. How many more miles did Ms. Webster fly on business this year? (Lesson 1.7)

   (A) 61,125 miles

   (B) 61,875 miles

   (C) 61,985 miles

   (D) 62,125 miles

Name _____

Remainders

Use counters to find the quotient and remainder.

1. 13 ÷ 4

    3 r1
    _____

2. 24 ÷ 7

    _____

3. 39 ÷ 5

    _____

4. 36 ÷ 8

    _____

5. 6)27

    _____

6. 25 ÷ 9

    _____

7. 3)17

    _____

8. 26 ÷ 4

    _____

Divide. Draw a quick picture to help.

9. 14 ÷ 3

10. 5)29

_____   _____

Problem Solving

11. What is the quotient and remainder in the
    division problem modeled below?

    _____

12. Mark drew the following model and said it
    represented the problem 21 ÷ 4. Is Mark's
    model correct? If so, what is the quotient
    and remainder? If not, what is the correct
    quotient and remainder?

    _____

    _____

    _____

Lesson Check

1. What is the quotient and remainder for 32 ÷ 6?

   (A) 4 r3

   (B) 5 r1

   (C) 5 r2

   (D) 6 r1

2. What is the remainder in the division problem modeled below?

   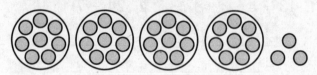

   (A) 8   (C) 3

   (B) 4   (D) 1

Spiral Review

3. Each kit to build a castle contains 235 parts. How many parts are in 4 of the kits? **(Lesson 2.6)**

   (A) 1,020

   (B) 940

   (C) 920

   (D) 840

4. In 2010, the population of Alaska was about 710,200. What is this number written in word form? **(Lesson 1.2)**

   (A) seven hundred ten thousand, two

   (B) seven hundred twelve thousand

   (C) seventy-one thousand, two

   (D) seven hundred ten thousand, two hundred

5. At the theater, one section of seats has 8 rows with 12 seats in each row. In the center of the first 3 rows are 4 broken seats that cannot be used. How many seats can be used in the section?

   (Lesson 2.9)

   (A) 84

   (B) 88

   (C) 92

   (D) 96

6. What partial products are shown by the model below? **(Lesson 3.4)**

   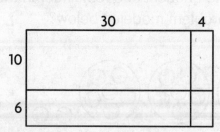

   (A) 300, 24

   (B) 300, 600, 40, 60

   (C) 300, 60, 40, 24

   (D) 300, 180, 40, 24

Name _____

Interpret the Remainder

Interpret the remainder to solve.

1. Hakeem has 100 tomato plants. He wants to plant them in rows of 8. How many full rows will he have?

Think: $100 \div 8$ is 12 with a remainder of 4. The question asks "how many full rows," so use only the quotient.

12 full rows
_____

2. A teacher has 27 students in her class. She asks the students to form as many groups of 4 as possible. How many students will not be in a group?

_____

3. A sporting goods company can ship 6 footballs in each carton. How many cartons are needed to ship 75 footballs?

_____

4. A carpenter has a board that is 10 feet long. He wants to make 6 table legs that are all the same length. What is the longest each leg can be?

_____

5. Allie wants to arrange her flower garden in 8 equal rows. She buys 60 plants. What is the greatest number of plants she can put in each row?

_____

Problem Solving REAL WORLD

6. Joanna has 70 beads. She uses 8 beads for each bracelet. She makes as many bracelets as possible. How many beads will Joanna have left over?

_____

7. A teacher wants to give 3 markers to each of her 25 students. Markers come in packages of 8. How many packages of markers will the teacher need?

_____

Lesson Check

1. Marcus sorts his 85 baseball cards into stacks of 9 cards each. How many stacks of 9 cards can Marcus make?

- Ⓐ 4
- Ⓑ 8
- Ⓒ 9
- Ⓓ 10

2. A minivan can hold up to 7 people. How many minivans are needed to take 45 people to a basketball game?

- Ⓐ 3
- Ⓑ 5
- Ⓒ 6
- Ⓓ 7

Spiral Review

3. Mrs. Wilkerson cut some oranges into 20 equal pieces to be shared by 6 friends. How many pieces did each person get and how many pieces were left over? **(Lesson 4.2)**

- Ⓐ 2 pieces with 4 pieces left over
- Ⓑ 3 pieces with 2 pieces left over
- Ⓒ 3 pieces with 4 pieces left over
- Ⓓ 4 pieces with 2 pieces left over

4. A school bought 32 new desks. Each desk cost $24. Which is the best estimate of how much the school spent on the new desks? **(Lesson 3.2)**

- Ⓐ $500
- Ⓑ $750
- Ⓒ $1,000
- Ⓓ $1,200

5. Kris has a box of 8 crayons. Sylvia's box has 6 times as many crayons as Kris's box. How many crayons are in Sylvia's box? **(Lesson 2.1)**

- Ⓐ 48
- Ⓑ 42
- Ⓒ 36
- Ⓓ 4

6. Yesterday, 1,743 people visited the fair. Today, there are 576 more people at the fair than yesterday. How many people are at the fair today? **(Lesson 1.8)**

- Ⓐ 1,167
- Ⓑ 2,219
- Ⓒ 2,319
- Ⓓ 2,367

Name _____

Divide Tens, Hundreds, and Thousands

Use basic facts and place value to find the quotient.

1. $3,600 \div 4 =$ __900__

Think: 3,600 is 36 hundreds.

Use the basic fact $36 \div 4 = 9$.

So, 36 hundreds $\div 4 = 9$ hundreds, or 900.

2. $240 \div 6 =$ _____

3. $5,400 \div 9 =$ _____

4. $300 \div 5 =$ _____

5. $4,800 \div 6 =$ _____

6. $420 \div 7 =$ _____

7. $150 \div 3 =$ _____

8. $6,300 \div 7 =$ _____

9. $1,200 \div 4 =$ _____

10. $360 \div 6 =$ _____

Find the quotient.

11. $28 \div 4 =$ _____

$280 \div 4 =$ _____

$2,800 \div 4 =$ _____

12. $18 \div 3 =$ _____

$180 \div 3 =$ _____

$1,800 \div 3 =$ _____

13. $45 \div 9 =$ _____

$450 \div 9 =$ _____

$4,500 \div 9 =$ _____

Problem Solving REAL WORLD

14. At an assembly, 180 students sit in 9 equal rows. How many students sit in each row?

_____

15. Hilary can read 560 words in 7 minutes. How many words can Hilary read in 1 minute?

_____

16. A company produces 7,200 gallons of bottled water each day. The company puts 8 one-gallon bottles in each carton. How many cartons are needed to hold all the one-gallon bottles produced in one day?

_____

17. An airplane flew 2,400 miles in 4 hours. If the plane flew the same number of miles each hour, how many miles did it fly in 1 hour?

_____

Lesson Check

1. A baseball player hits a ball 360 feet to the outfield. It takes the ball 4 seconds to travel this distance. How many feet does the ball travel in 1 second?

   (A) 9 feet

   (B) 40 feet

   (C) 90 feet

   (D) 900 feet

2. Sebastian rides his bike 2,000 meters in 5 minutes. How many meters does he bike in 1 minute?

   (A)   4 meters

   (B) 40 meters

   (C) 50 meters

   (D) 400 meters

Spiral Review

3. A full container of juice holds 64 ounces. How many 7-ounce servings of juice are in a full container? (Lesson 4.3)

   (A) 1

   (B) 8

   (C) 9

   (D) 10

4. Paolo pays $244 for 5 identical calculators. Which is the best estimate of how much Paolo pays for one calculator?
   (Lesson 4.1)

   (A) $40

   (B) $50

   (C) $60

   (D) $245

5. A football team paid $28 per jersey. They bought 16 jerseys. How much money did the team spend on jerseys? (Lesson 3.5)

   (A) $44

   (B) $196

   (C) $408

   (D) $448

6. Suzanne bought 50 apples at the apple orchard. She bought 4 times as many red apples as green apples. How many more red apples than green apples did Suzanne buy? (Lesson 2.2)

   (A) 10

   (B) 25

   (C) 30

   (D) 40

Name _____

Estimate Quotients Using Compatible Numbers

Use compatible numbers to estimate the quotient.

1. 389 ÷ 4

400 ÷ 4 = 100

2. 358 ÷ 3

3. 784 ÷ 8

4. 179 ÷ 9

5. 315 ÷ 8

6. 2,116 ÷ 7

7. 4,156 ÷ 7

8. 474 ÷ 9

Use compatible numbers to find two estimates that the quotient is between.

9. 1,624 ÷ 3

10. 2,593 ÷ 6

11. 1,045 ÷ 2

12. 1,754 ÷ 9

13. 2,363 ÷ 8

14. 1,649 ÷ 5

15. 5,535 ÷ 7

16. 3,640 ÷ 6

Problem Solving REAL WORLD

17. A CD store sold 3,467 CDs in 7 days. About the same number of CDs were sold each day. About how many CDs did the store sell each day?

18. Marcus has 731 books. He puts about the same number of books on each of 9 shelves in his a bookcase. About how many books are on each shelf?

Lesson Check

1. Jamal is planting seeds for a garden nursery. He plants 9 seeds in each container. If Jamal has 296 seeds to plant, about how many containers will he use?

   (A) about 20
   (B) about 30
   (C) about 200
   (D) about 300

2. Winona purchased a set of vintage beads. There are 2,140 beads in the set. If she uses the beads to make bracelets that have 7 beads each, about how many bracelets can she make?

   (A) about 30
   (B) about 140
   (C) about 300
   (D) about 14,000

Spiral Review

3. A train traveled 360 miles in 6 hours. How many miles per hour did the train travel? (Lesson 4.4)

   (A) 60 miles per hour
   (B) 66 miles per hour
   (C) 70 miles per hour
   (D) 600 miles per hour

4. An orchard has 12 rows of pear trees. Each row has 15 pear trees. How many pear trees are there in the orchard? (Lesson 3.6)

   (A) 170
   (B) 180
   (C) 185
   (D) 190

5. Megan rounded 366,458 to 370,000. To which place did Megan round the number? (Lesson 1.4)

   (A) hundred thousands
   (B) ten thousands
   (C) thousands
   (D) hundreds

6. Mr. Jessup, an airline pilot, flies 1,350 miles a day. How many miles will he fly in 8 days? (Lesson 2.11)

   (A) 1,358 miles
   (B) 8,400 miles
   (C) 10,800 miles
   (D) 13,508 miles

Name _____

Division and the Distributive Property

Find the quotient.

1. $54 \div 3 = ($ ___**30**___ $\div 3) + ($ ___**24**___ $\div 3)$

       $= $ ___**10**___ $+$ ___**8**___

       $= $ ___**18**___

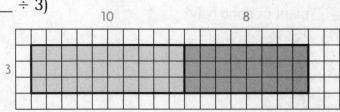

2. $81 \div 3 = $ _____

3. $232 \div 4 = $ _____

4. $305 \div 5 = $ _____

5. $246 \div 6 = $ _____

6. $69 \div 3 = $ _____

7. $477 \div 9 = $ _____

8. $224 \div 7 = $ _____

9. $72 \div 4 = $ _____

10. $315 \div 3 = $ _____

Problem Solving REAL WORLD

11. Cecily picked 219 apples. She divided the apples equally into 3 baskets. How many apples are in each basket?

_____

12. Jordan has 260 basketball cards. He divides them into 4 equal groups. How many cards are in each group?

_____

13. The Wilsons drove 324 miles in 6 hours. If they drove the same number of miles each hour, how many miles did they drive in 1 hour?

_____

14. Phil has 189 stamps to put into his stamp album. He puts the same number of stamps on each of 9 pages. How many stamps does Phil put on each page?

_____

Lesson Check

1. A landscaping company planted 176 trees in 8 equal rows in the new park. How many trees did the company plant in each row?

   (A) 18

   (B) 20

   (C) 22

   (D) 24

2. Arnold can do 65 pushups in 5 minutes. How many pushups can he do in 1 minute?

   (A) 11

   (B) 13

   (C) 15

   (D) 17

Spiral Review

3. Last Saturday, there were 1,486 people at the Cineplex. There were about the same number of people in each of the 6 theaters. Which is the best estimate of the number of people in each theater?
   (Lesson 4.5)

   (A) between 20 and 30

   (B) between 80 and 90

   (C) between 100 and 200

   (D) between 200 and 300

4. Nancy walked 50 minutes each day for 4 days last week. Gillian walked 35 minutes each day for 6 days last week. Which statement is true?
   (Lesson 3.7)

   (A) Gillian walked 10 minutes more than Nancy.

   (B) Gillian walked 20 minutes more than Nancy.

   (C) Nancy walked 10 minutes more than Gillian.

   (D) Nancy walked 15 minutes more than Gillian.

5. Three boys share 28 toy cars equally. Which best describes how the cars are shared? (Lesson 4.2)

   (A) Each gets 3 cars with 1 left over.

   (B) Each gets 8 cars with 2 left over.

   (C) Each gets 9 cars with 1 left over.

   (D) Each gets 10 cars with 2 left over.

6. An airplane flies at a speed of 474 miles per hour. How many miles does the plane fly in 5 hours? (Lesson 2.11)

   (A) 2,070 miles

   (B) 2,140 miles

   (C) 2,370 miles

   (D) 2,730 miles

Name _____

Divide Using Repeated Subtraction

Use repeated subtraction to divide.

1. $42 \div 3 =$ **14** 2. $72 \div 4 =$ _____ 3. $93 \div 3 =$ _____

$$
\begin{array}{r}
3\overline{)42} \\
-30 \leftarrow 10 \times 3 \quad 10 \\
\hline
12 \\
-12 \leftarrow 4 \times 3 \quad +4 \\
\hline
0 \qquad\qquad 14
\end{array}
$$

4. $35 \div 4$ _____ 5. $93 \div 10$ _____ 6. $86 \div 9$ _____

Draw a number line to divide.

7. $70 \div 5 =$ _____

Problem Solving REAL WORLD

8. Gretchen has 48 small shells. She uses 2 shells to make one pair of earrings. How many pairs of earrings can she make?

9. James wants to purchase a telescope for $54. If he saves $3 per week, in how many weeks will he have saved enough to purchase the telescope?

_____ _____

Lesson Check

1. Randall collects postcards that his friends send him when they travel. He can put 6 cards on one scrapbook page. How many pages does Randall need to fit 42 postcards?

    (A) 3

    (B) 4

    (C) 6

    (D) 7

2. Ari stocks shelves at a grocery store. He puts 35 cans of juice on each shelf. The shelf has 4 equal rows and another row with only 3 cans. How many cans are in each of the equal rows?

    (A) 6

    (B) 7

    (C) 8

    (D) 9

Spiral Review

3. Fiona sorted her CDs into separate bins. She placed 4 CDs in each bin. If she has 160 CDs, how many bins did she fill? (Lesson 4.4)

    (A) 4

    (B) 16

    (C) 40

    (D) 156

4. Eamon is arranging 39 books on 3 shelves. If he puts the same number of books on each shelf, how many books will there be on each shelf? (Lesson 4.6)

    (A) 11

    (B) 12

    (C) 13

    (D) 14

5. A newborn boa constrictor measures 18 inches long. An adult boa constrictor measures 9 times the length of the newborn plus 2 inches. How long is the adult? (Lesson 2.12)

    (A) 142 inches

    (B) 162 inches

    (C) 164 inches

    (D) 172 inches

6. Madison has 6 rolls of coins. Each roll has 20 coins. How many coins does Madison have in all? (Lesson 2.3)

    (A) 110

    (B) 120

    (C) 125

    (D) 130

Divide Using Partial Quotients

Divide. Use partial quotients.

1. 8)184
   -80   10 × 8   10
   ___
   104
   -80   10 × 8   10
   ___
   24
   -24   3 × 8   +3
   ___
   0   23

2. 6)258

3. 5)630

Divide. Use rectangular models to record the partial quotients.

4. 246 ÷ 3 = _____

5. 126 ÷ 2 = _____

6. 605 ÷ 5 = _____

Divide. Use either way to record the partial quotients.

7. 492 ÷ 3 = _____

8. 224 ÷ 7 = _____

9. 692 ÷ 4 = _____

Problem Solving  REAL WORLD

10. Allison took 112 photos on vacation. She wants to put them in a photo album that holds 4 photos on each page. How many pages can she fill?

_____

11. Hector saved $726 in 6 months. He saved the same amount each month. How much did Hector save each month?

_____

Lesson Check

1. Annaka used partial quotients to divide 145 ÷ 5. Which shows a possible sum of partial quotients?

   Ⓐ 50 + 50 + 45

   Ⓑ 100 + 40 + 5

   Ⓒ 10 + 10 + 9

   Ⓓ 10 + 4 + 5

2. Mel used partial quotients to find the quotient 378 ÷ 3. Which might show the partial quotients that Mel found?

   Ⓐ 100, 10, 10, 9

   Ⓑ 100, 10, 10, 6

   Ⓒ 100, 30, 30, 6

   Ⓓ 300, 70, 8

Spiral Review

3. What are the partial products of 42 × 5? (Lesson 2.7)

   Ⓐ 9 and 7

   Ⓑ 20 and 10

   Ⓒ 200 and 7

   Ⓓ 200 and 10

4. Mr. Watson buys 4 gallons of paint that cost $34 per gallon. How much does Mr. Watson spend on paint? (Lesson 2.10)

   Ⓐ $38

   Ⓑ $126

   Ⓒ $136

   Ⓓ $1,216

5. Use the area model to find the product 28 × 32. (Lesson 3.3)

   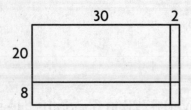

   Ⓐ 840

   Ⓑ 856

   Ⓒ 880

   Ⓓ 896

6. An adult male lion eats about 108 pounds of meat per week. About how much meat does an adult male lion eat in one day? (Lesson 4.1)

   Ⓐ about 14 pounds

   Ⓑ about 15 pounds

   Ⓒ about 16 pounds

   Ⓓ about 17 pounds

Name _____

Model Division with Regrouping

Divide. Use base-ten blocks.

1. 63 ÷ 4 <u>**15 r3**</u>

2. 83 ÷ 3 _____

Divide. Draw quick pictures. Record the steps.

3. 85 ÷ 5 _____

4. 97 ÷ 4 _____

Problem Solving REAL WORLD

5. Tamara sold 92 cold drinks during her 2-hour shift at a festival food stand. If she sold the same number of drinks each hour, how many cold drinks did she sell each hour?

_____

6. In 3 days Donald earned $42 running errands. He earned the same amount each day. How much did Donald earn from running errands each day?

_____

Lesson Check

1. Gail bought 80 buttons to put on the shirts she makes. She uses 5 buttons for each shirt. How many shirts can Gail make with the buttons she bought?

    (A) 14

    (B) 16

    (C) 17

    (D) 18

2. Marty counted how many breaths he took in 3 minutes. In that time, he took 51 breaths. He took the same number of breaths each minute. How many breaths did Marty take in one minute?

    (A) 15

    (B) 16

    (C) 17

    (D) 19

Spiral Review

3. Kate is solving brain teasers. She solved 6 brain teasers in 72 minutes. How long did she spend on each brain teaser? (Lesson 4.7)

    (A) 12 minutes

    (B) 14 minutes

    (C) 18 minutes

    (D) 22 minutes

4. Jenny works at a package delivery store. She puts mailing stickers on packages. Each package needs 5 stickers. How many stickers will Jenny use if she is mailing 105 packages? (Lesson 2.11)

    (A) 725   (C) 525

    (B) 625   (D) 21

5. The Puzzle Company packs standard-sized puzzles into boxes that hold 8 puzzles. How many boxes would it take to pack up 192 standard-sized puzzles? (Lesson 4.6)

    (A) 12

    (B) 16

    (C) 22

    (D) 24

6. Mt. Whitney in California is 14,494 feet tall. Mt. McKinley in Alaska is 5,826 feet taller than Mt. Whitney. How tall is Mt. McKinley? (Lesson 1.6)

    (A) 21,310 feet

    (B) 20,320 feet

    (C) 20,230 feet

    (D) 19,310 feet

Place the First Digit

Divide.

1. $\begin{array}{r} 62 \\ 3)\overline{186} \\ -18\downarrow \\ \hline 06 \\ -6 \\ \hline 0 \end{array}$

2. $4)\overline{298}$

3. $3)\overline{461}$

4. $9)\overline{315}$

5. $2)\overline{766}$

6. $4)\overline{604}$

7. $6)\overline{796}$

8. $5)\overline{449}$

9. $6)\overline{756}$

10. $7)\overline{521}$

11. $5)\overline{675}$

12. $8)\overline{933}$

Problem Solving

13. There are 132 projects in the science fair. If 8 projects can fit in a row, how many full rows of projects can be made? How many projects are in the row that is not full?

14. There are 798 calories in six 10-ounce bottles of apple juice. How many calories are there in one 10-ounce bottle of apple juice?

_____   _____

Lesson Check

1. To divide $572 \div 4$, Stanley estimated to place the first digit of the quotient. In which place is the first digit of the quotient?

   Ⓐ ones

   Ⓑ tens

   Ⓒ hundreds

   Ⓓ thousands

2. Onetta biked 325 miles in 5 days. If she biked the same number of miles each day, how far did she bike each day?

   Ⓐ 1,625 miles

   Ⓑ 320 miles

   Ⓒ   65 miles

   Ⓓ   61 miles

Spiral Review

3. Mort makes beaded necklaces that he sells for $32 each. About how much will Mort make if he sells 36 necklaces at the local art fair? (Lesson 3.2)

   Ⓐ $120

   Ⓑ $900

   Ⓒ $1,200

   Ⓓ $1,600

4. Which is the best estimate of 54×68? (Lesson 3.2)

   Ⓐ 4,200

   Ⓑ 3,500

   Ⓒ 3,000

   Ⓓ 350

5. Ms. Eisner pays $888 for 6 nights in a hotel. How much does Ms. Eisner pay per night? (Lesson 4.8)

   Ⓐ $5,328

   Ⓑ $882

   Ⓒ $148

   Ⓓ $114

6. Which division problem does the model show? (Lesson 4.9)

   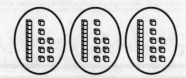

   Ⓐ $42 \div 3$ Ⓒ $51 \div 3$

   Ⓑ $44 \div 3$ Ⓓ $54 \div 3$

Name _____

Divide by 1-Digit Numbers

Divide and check.

1.
$$
\begin{array}{r}
318 \\
2\overline{)636} \\
-6 \downarrow \\
\overline{03} \\
-2 \downarrow \\
\overline{16} \\
-16 \\
\overline{0}
\end{array}
\qquad
\begin{array}{r}
318 \\
\times\ 2 \\
\hline
636
\end{array}
$$

2. $4\overline{)631}$

3. $8\overline{)906}$

4. $6\overline{)6,739}$

5. $4\overline{)2,328}$

6. $5\overline{)7,549}$

Problem Solving REAL WORLD

Use the table for 7 and 8.

7. The Briggs rented a car for 5 weeks. What was the cost of their rental car per week?

_____

8. The Lees rented a car for 4 weeks. The Santos rented a car for 2 weeks. Whose weekly rental cost was lower? **Explain**.

_____

_____

Rental Car Costs	
Family	Total Cost
Lee	$632
Brigg	$985
Santo	$328

Lesson Check

1. Which expression can be used to check the quotient 646 ÷ 3?

    (A) (251 × 3) + 1

    (B) (215 × 3) + 2

    (C) (215 × 3) + 1

    (D) 646 × 3

2. There are 8 volunteers at the telethon. The goal for the evening is to raise $952. If each volunteer raises the same amount, what is the minimum amount each needs to raise to meet the goal?

    (A) $7,616

    (B) $944

    (C) $119

    (D) $106

Spiral Review

3. Which product is shown by the model?
    (Lesson 2.5)

    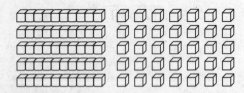

    (A) 5 × 15 = 75

    (B) 5 × 16 = 80

    (C) 5 × 17 = 75

    (D) 5 × 17 = 85

4. The computer lab at a high school ordered 26 packages of CDs. There were 50 CDs in each package. How many CDs did the computer lab order? (Lesson 3.1)

    (A) 1,330

    (B) 1,300

    (C) 1,030

    (D) 130

5. Which of the following division problems has a quotient with the first digit in the hundreds place? (Lesson 4.10)

    (A) 892 ÷ 9

    (B) 644 ÷ 8

    (C) 429 ÷ 5

    (D) 306 ÷ 2

6. Sharon has 64 ounces of juice. She is going to use the juice to fill as many 6-ounce glasses as possible. She will drink the leftover juice. How much juice will Sharon drink? (Lesson 4.3)

    (A) 4 ounces

    (B) 6 ounces

    (C) 10 ounces

    (D) 12 ounces

Name _____

Problem Solving • Multistep Division Problems

Solve. Draw a diagram to help you.

1. There are 3 trays of eggs. Each tray holds 30 eggs. How many people can be served if each person eats 2 eggs?

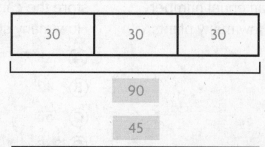

Multiply to find the total number of eggs.

Think: What do I need to find? How can I draw a diagram to help?

Divide to find how many people can be served 2 eggs.

__45 people can be served.__

2. There are 8 pencils in a package. How many packages will be needed for 28 children if each child gets 4 pencils?

_____

3. There are 3 boxes of tangerines. Each box has 93 tangerines. The tangerines will be divided equally among 9 classrooms. How many tangerines will each classroom get?

_____

4. Misty has 84 photos from her vacation and 48 photos from a class outing. She wants to put all the photos in an album with 4 photos on each page. How many pages does she need?

_____

Lesson Check

1. Gavin buys 89 blue pansies and 86 yellow pansies. He will plant the flowers in 5 rows with an equal number of plants in each row. How many plants will be in each row?

   (A) 875

   (B) 175

   (C) 35

   (D) 3

2. A pet store receives 7 boxes of cat food. Each box has 48 cans. The store wants to store the cans in equal stacks of 8 cans. How many stacks can be formed?

   (A) 8

   (B) 42

   (C) 56

   (D) 336

Spiral Review

3. What product does the model show? (Lesson 3.4)

   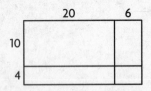

   (A) 284

   (B) 304

   (C) 340

   (D) 364

4. Mr. Hatch bought 4 round-trip airplane tickets for $417 each. He also paid $50 in baggage fees. How much did Mr. Hatch spend? (Lesson 2.12)

   (A) $467

   (B) $1,698

   (C) $1,718

   (D) $16,478

5. Mae read 976 pages in 8 weeks. She read the same number of pages each week. How many pages did she read each week? (Lesson 4.10)

   (A) 109

   (B) 120

   (C) 122

   (D) 984

6. Yolanda and her 3 brothers shared a box of 156 toy dinosaurs. About how many dinosaurs did each child get? (Lesson 4.5)

   (A) 40

   (B) 50

   (C) 60

   (D) 80

Chapter 4 Extra Practice

Lessons 4.1, 4.5

Estimate the quotient.

1. $67 \div 4$ **2.** $72 \div 5$ **3.** $213 \div 3$ **4.** $484 \div 6$

_____   _____   _____   _____

5. $446 \div 7$ **6.** $1,246 \div 4$ **7.** $708 \div 9$ **8.** $2,657 \div 3$

_____   _____   _____   _____

Lesson 4.2

Use counters or quick pictures to find the quotient and remainder.

1. $44 \div 5$ **2.** $8\overline{)21}$ **3.** $4\overline{)75}$ **4.** $76 \div 6$

_____   _____   _____   _____

Lesson 4.3

Interpret the remainder to solve.

1. Kelly divides 29 markers equally among 7 friends. If Kelly keeps the leftover markers, how many markers will she keep?

2. Dave has a board that is 29 inches long. He cuts the board into 4 equal pieces. How long will each piece be?

_____   _____

3. Eight students can ride in each van. How many vans are needed for 29 students?

4. Mac has 40 ounces of juice. He pours 6 ounces in each glass. How many glasses can he fill?

_____   _____

Lesson 4.4

Use basic facts and place value to find the quotient.

1. $120 \div 4 =$ _____   **2.** $280 \div 7 =$ _____   **3.** $3,000 \div 5 =$ _____

4. $4,800 \div 6 =$ _____   **5.** $5,600 \div 8 =$ _____   **6.** $6,300 \div 9 =$ _____

Lessons 4.6 - 4.7

Choose a method and divide.

1. 68 ÷ 4 _____

2. 48 ÷ 3 _____

3. 108 ÷ 9 _____

4. 74 ÷ 2 _____

5. 122 ÷ 5 _____

6. 165 ÷ 6 _____

Lessons 4.8 - 4.9

Divide.

1. 4)848

2. 7)287

3. 5)405

4. 3)696

5. 96 ÷ 6 _____

6. 76 ÷ 5 _____

7. 58 ÷ 4 _____

8. 85 ÷ 2 _____

Lessons 4.10 - 4.11

Divide and check.

1. 4)896

2. 5)833

3. 6)527

4. 3)935

5. 8)1,976

6. 6)1,042

Lesson 4.12

Solve. Draw a diagram to help you.

1. Ellis has 2 dozen white baseballs and 4 dozen yellow baseballs. He needs to divide them into cartons that hold 6 each. How many cartons can he fill?

2. A family of 2 adults and 3 children went out to dinner. The total bill was $42. Each child's dinner cost $4. How much did each adult's dinner cost?

_____ _____

Chapter 5 · School-Home Letter

Vocabulary

common factor A number that is a factor of two or more numbers

common multiple A number that is a multiple of two or more numbers

divisible A number is divisible by another number if the quotient is a counting number and the remainder is zero.

composite number A whole number greater than 1 that has more than two factors

prime number A number that has exactly two factors: 1 and itself

Dear Family,

Throughout the next few weeks, our math class will be working with factors, multiples, and patterns. The students will study and learn to find factors and multiples and work with number patterns.

Here is a sample of how your child will be taught.

🔑 MODEL Find Factor Pairs

Use division to find all the factor pairs for 36.
Divisibility rules can help.

Factors of 36		Divisibility Rules
$36 \div 1 = 36$	1, 36	Every whole number is divisible by 1.
$36 \div 2 = 18$	2, 18	The number is even. It's divisible by 2.
$36 \div 3 = 12$	3, 12	The sum of the digits is divisible by 3.
$36 \div 6 = 6$	6, 6	The number is even, and divisible by 3.
$36 \div 9 = 4$	9, 4	The sum of the digits is divisible by 9.

Tips

Divisibility
A whole number is divisible by another whole number when the quotient is a whole number and the remainder is 0.

Activity

Using the divisibility rules, have your child find all the factor pairs for these numbers:
18, 48, 39, 63

© Houghton Mifflin Harcourt Publishing Company

Carta para la casa

© Houghton Mifflin Harcourt Publishing Company

Vocabulario

factor común Un número que es factor de dos o más números

común múltiplo Un número que es múltiplo de dos o más números

divisible Un número es divisible entre otro número si el cociente es un número entero y el residuo es cero.

número compuesto Un número entero mayor que 1 que tiene más de dos factores

número primo Un número que tiene exactamente dos factores: 1 y él mismo

Querida familia,

Durante las próximas semanas, en la clase de matemáticas trabajaremos con factores, múltiplos y patrones. Aprenderemos a hallar factores y múltiplos y a trabajar con patrones de números.

Este es un ejemplo de la manera como aprenderemos.

🔑 MODELO   Hallar pares de factores

Usa la división para hallar todos los pares de factores para 36.
Las reglas de divisibilidad te pueden ayudar.

Factores de 36		Reglas de divisibilidad
$36 \div 1 = 36$	1, 36	Todos los números enteros son divisibles entre 1.
$36 \div 2 = 18$	2, 18	El número es par. Es divisible entre 2.
$36 \div 3 = 12$	3, 12	La suma de los dígitos es divisible entre 3.
$36 \div 6 = 6$	6, 6	El número es par y divisible entre 3.
$36 \div 9 = 4$	9, 4	La suma de los dígitos es divisible entre 9.

Pistas

Divisibilidad

Un número entero es divisible entre otro número entero si el cociente es un número entero y el residuo es 0.

Actividad

Usando las reglas de divisibilidad, pida a su niño o niña que halle todos los pares de factores para estos números: 18, 48, 39, 63.

Name _____

Model Factors

Use tiles to find all the factors of the product.
Record the arrays on grid paper and write the factors shown.

1. 15

$1 \times 15 = 15$

$3 \times 5 = 15$

1, 3, 5, 15

2. 30

_____

_____

3. 45

_____

_____

4. 19

_____

_____

5. 40

_____

_____

_____

6. 36

_____

_____

_____

7. 22

_____

_____

_____

8. 4

_____

_____

_____

9. 26

_____

_____

10. 49

_____

_____

11. 32

_____

_____

12. 23

_____

_____

Problem Solving REAL WORLD

13. Brooke has to set up 70 chairs in equal rows for the class talent show. But, there is not room for more than 20 rows. What are the possible number of rows that Brooke could set up?

_____

14. Eduardo thinks of a number between 1 and 20 that has exactly 5 factors. What number is he thinking of?

_____

Lesson Check

1. Which of the following lists all the factors of 24?

   Ⓐ 1, 4, 6, 24

   Ⓑ 1, 3, 8, 24

   Ⓒ 3, 4, 6, 8

   Ⓓ 1, 2, 3, 4, 6, 8, 12, 24

2. Natalia has 48 tiles. Which of the following shows a factor pair for the number 48?

   Ⓐ 4 and 8

   Ⓑ 6 and 8

   Ⓒ 2 and 12

   Ⓓ 3 and 24

Spiral Review

3. The Pumpkin Patch is open every day. If it sells 2,750 pounds of pumpkins each day, about how many pounds does it sell in 7 days? (Lesson 2.4)

   Ⓐ   210 pounds

   Ⓑ   2,100 pounds

   Ⓒ 14,000 pounds

   Ⓓ 21,000 pounds

4. What is the remainder in the division problem modeled below? (Lesson 4.2)

   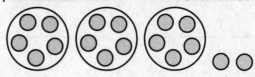

   Ⓐ 2   Ⓒ 5

   Ⓑ 3   Ⓓ 17

5. Which number sentence is represented by the following array? (Lesson 2.1)

   Ⓐ 4 × 5 = 20

   Ⓑ 4 × 4 = 16

   Ⓒ 5 × 2 = 10

   Ⓓ 5 × 5 = 25

6. Channing jogs 10 miles a week. How many miles will she jog in 52 weeks?
   (Lesson 3.1)

   Ⓐ   30 miles

   Ⓑ 120 miles

   Ⓒ 200 miles

   Ⓓ 520 miles

Factors and Divisibility

Is 6 a factor of the number? Write _yes_ or _no_.

1. 36   **2.** 56   **3.** 42   **4.** 66

Think: $6 \times 6 = 36$

_____ **yes** _____   _____   _____   _____

Is 5 a factor of the number? Write _yes_ or _no_.

5. 38   **6.** 45   **7.** 60   **8.** 39

_____   _____   _____   _____

List all the factor pairs in the table.

9.

Factors of 12	
_____ × _____ = _____	_____, _____
_____ × _____ = _____	_____, _____
_____ × _____ = _____	_____, _____

10.

Factors of 25	
_____ × _____ = _____	_____, _____
_____ × _____ = _____	_____, _____

11. List all the factor pairs for 48. Make a table to help.

_____

_____

Problem Solving  REAL WORLD

12. Bryson buys a bag of 64 plastic miniature dinosaurs. Could he distribute them equally into six storage containers and not have any left over? **Explain**.

_____

13. Lori wants to distribute 35 peaches equally into baskets. She will use more than 1 but fewer than 10 baskets. How many baskets does Lori need?

_____

Lesson Check

1. Which of the following numbers has 9 as a factor?

   (A) 28

   (B) 30

   (C) 39

   (D) 45

2. Which of the following numbers does NOT have 5 as a factor?

   (A) 15

   (B) 28

   (C) 30

   (D) 45

Spiral Review

3. Which of the following shows a strategy to use to find 4×275? (Lesson 2.8)

   (A) $(4 \times 300) + (4 \times 25)$

   (B) $(4 \times 300) - (4 \times 25)$

   (C) $(4 \times 275) - 100$

   (D) $(4 \times 200) + 75$

4. Jack broke apart 5×216 as $(5 \times 200) + (5 \times 16)$ to multiply mentally. What strategy did Jack use? (Lesson 2.8)

   (A) the Commutative Property

   (B) the Associative Property

   (C) halving and doubling

   (D) the Distributive Property

5. Jordan has $55. She earns $67 by doing chores. How much money does Jordan have now? (Lesson 1.6)

   (A) $122

   (B) $130

   (C) $112

   (D) $12

6. Trina has 72 collector's stamps. She puts 43 of the stamps into a stamp book. How many stamps are left? (Lesson 1.7)

   (A) 29

   (B) 31

   (C) 39

   (D) 115

Problem Solving • Common Factors

Solve each problem.

1. Grace is preparing grab bags for her store's open house. She has 24 candles, 16 pens, and 40 figurines. Each grab bag will have the same number of items, and all the items in a bag will be the same. How many items can Grace put in each bag?

Find the common factors of 24, 16, and 40.

1, 2, 4, or 8 items

2. Simon is making wreaths to sell. He has 60 bows, 36 silk roses, and 48 silk carnations. He wants to put the same number of items on each wreath. All the items on a wreath will be the same type. How many items can Simon put on each wreath?

_____

3. Justin has 20 pencils, 25 erasers, and 40 paper clips. He organizes them into groups with the same number of items in each group. All the items in a group will be the same type. How many items can he put in each group?

_____

4. A food bank has 50 cans of vegetables, 30 loaves of bread, and 100 bottles of water. The volunteers will put the items into boxes. Each box will have the same number of food items and all the items in the box will be the same type. How many items can they put in each box?

_____

5. A debate competition has participants from three different schools: 15 from James Elementary, 18 from George Washington School, and 12 from the MLK Jr. Academy. All teams must have the same number of students. Each team can have only students from the same school. How many students can be on each team?

_____

Lesson Check

1. What are all the common factors of 24, 64, and 88?

   (A) 1 and 4

   (B) 1, 4, and 8

   (C) 1, 4, 8, and 12

   (D) 1, 4, 8, and 44

2. Which number is NOT a common factor of 15, 45, and 90?

   (A) 3

   (B) 5

   (C) 10

   (D) 15

Spiral Review

3. Dan puts $11 of his allowance in his savings account every week. How much money will he have after 15 weeks? (Lesson 3.4)

   (A) $165

   (B) $132

   (C) $110

   (D) $26

4. James is reading a book that is 1,400 pages. He will read the same number of pages each day. If he reads the book in 7 days, how many pages will he read each day? (Lesson 4.4)

   (A) 20

   (B) 50

   (C) 140

   (D) 200

5. Emma volunteered at an animal shelter for a total of 119 hours over 6 weeks. Which is the best estimate of the number of hours she volunteered each week? (Lesson 4.5)

   (A) 10 hours

   (B) 20 hours

   (C) 120 hours

   (D) 714 hours

6. Which strategy can be used to multiply 6×198 mentally? (Lesson 2.8)

   (A) $6 \times 198 = (6 \times 19) + (6 \times 8)$

   (B) $6 \times 198 = (6 \times 200) + (6 \times 2)$

   (C) $6 \times 198 = (6 \times 200) - (6 \times 2)$

   (D) $6 \times 198 = (6 + 200) \times (6 + 2)$

Name _____

Factors and Multiples

Is the number a multiple of 8? Write *yes* or *no*.

1. 4   **2.** 8   **3.** 20   **4.** 40

Think: Since $4 \times 2 = 8$,
4 is a *factor* of 8, not a
multiple of 8.

       no
_____ _____ _____ _____

List the next nine multiples of each number.
Find the common multiples.

5. Multiples of 4: 4, _____

     Multiples of 7: 7, _____

     Common multiples: _____

6. Multiples of 3: 3, _____

     Multiples of 9: 9, _____

     Common multiples: _____

7. Multiples of 6: 6, _____

     Multiples of 8: 8, _____

     Common multiples: _____

Tell whether 24 is a factor or multiple of the number.
Write *factor*, *multiple*, or *neither*.

8. 6 _____   **9.** 36 _____   **10.** 48 _____

Problem Solving REAL WORLD

11. Ken paid $12 for two magazines. The cost of each magazine was a multiple of $3. What are the possible prices of the magazines?

12. Jodie bought some shirts for $6 each. Marge bought some shirts for $8 each. The girls spent the same amount of money on shirts. What is the least amount they could have spent?

_____   _____

Lesson Check

1. Which list shows numbers that are all multiples of 4?

   Ⓐ 2, 4, 6, 8

   Ⓑ 3, 7, 11, 15, 19

   Ⓒ 4, 14, 24, 34

   Ⓓ 4, 8, 12, 16

2. Which of the following numbers is a common multiple of 5 and 9?

   Ⓐ 9

   Ⓑ 14

   Ⓒ 36

   Ⓓ 45

Spiral Review

3. Jenny has 50 square tiles. She arranges the tiles into a rectangular array of 4 rows. How many tiles will be left over? **(Lesson 4.3)**

   Ⓐ 0

   Ⓑ 1

   Ⓒ 2

   Ⓓ 4

4. Jerome added two numbers. The sum was 83. One of the numbers was 45. What was the other number? **(Lesson 1.7)**

   Ⓐ 38

   Ⓑ 48

   Ⓒ 42

   Ⓓ 128

5. There are 18 rows of seats in the auditorium. There are 24 seats in each row. How many seats are in the auditorium in all? **(Lesson 3.5)**

   Ⓐ 42

   Ⓑ 108

   Ⓒ 412

   Ⓓ 432

6. The population of Riverdale is 6,735. What is the value of the 7 in the number 6,735? **(Lesson 1.2)**

   Ⓐ 7

   Ⓑ 700

   Ⓒ 735

   Ⓓ 7,000

Name _____

Prime and Composite Numbers

Tell whether the number is *prime* or *composite*.

1. 47

Think: Does 47 have other factors besides 1 and itself?

prime
_____

2. 68

_____

3. 52

_____

4. 63

_____

5. 75

_____

6. 31

_____

7. 77

_____

8. 59

_____

9. 87

_____

10. 72

_____

11. 49

_____

12. 73

_____

Problem Solving  REAL WORLD

13. Kai wrote the number 85 on the board. Is 85 prime or composite? **Explain.**

_____

_____

_____

14. Lisa says that 43 is a 2-digit odd number that is composite. Is she correct? **Explain.**

_____

_____

_____

Lesson Check

1. The number 5 is:

   (A) prime

   (B) composite

   (C) both prime and composite

   (D) neither prime nor composite

2. The number 1 is:

   (A) prime

   (B) composite

   (C) both prime and composite

   (D) neither prime nor composite

Spiral Review

3. A recipe for a vegetable dish contains a total of 924 calories. The dish serves 6 people. How many calories are in each serving? (Lesson 4.10)

   (A) 134 calories

   (B) 150 calories

   (C) 154 calories

   (D) 231 calories

4. A store clerk has 45 shirts to pack in boxes. Each box holds 6 shirts. What is the fewest boxes the clerk will need to pack all the shirts? (Lesson 4.3)

   (A) 9

   (B) 8

   (C) 7

   (D) 6

5. Which number rounds to 200,000? (Lesson 1.4)

   (A) 289,005

   (B) 251,659

   (C) 152,909

   (D) 149,889

6. What is the word form of the number 602,107? (Lesson 1.2)

   (A) six hundred twenty thousand, seventeen

   (B) six hundred two thousand, one hundred seven

   (C) six hundred twenty-one thousand, seventeen

   (D) six hundred two thousand, one hundred seventy

Name _____

Number Patterns

Use the rule to write the first twelve numbers in the pattern.
Describe another pattern in the numbers.

1. Rule: *Add 8.*   First term: 5

Think: Add 8.

5   13   21   29   37

5, 13, 21, 29, 37, 45, 53, 61, 69, 77, 85, 93

All the terms are odd numbers.

2. Rule: *Subtract 7.*   First term: 95

_____

_____

3. Rule: *Add 15, subtract 10.*   First term: 4

_____

_____

4. Rule: *Add 1, multiply by 2.*   First term: 2

_____

_____

Problem Solving

5. Barb is making a bead necklace. She
strings 1 white bead, then 3 blue beads,
then 1 white bead, and so on. Write the
numbers for the first eight beads that are
white. What is the rule for the pattern?

_____

_____

6. An artist is arranging tiles in rows to
decorate a wall. Each new row has 2 fewer
tiles than the row below it. If the first row
has 23 tiles, how many tiles will be in the
seventh row?

_____

_____

Lesson Check

1. The rule for a pattern is *add 6*. The first term is 5. Which of the following numbers is a term in the pattern?

   (A) 6

   (B) 12

   (C) 17

   (D) 22

2. What are the next two terms in the pattern 3, 6, 5, 10, 9, 18, 17, . . .?

   (A) 16, 15

   (B) 30, 31

   (C) 33, 34

   (D) 34, 33

Spiral Review

3. To win a game, Roger needs to score 2,000 points. So far, he has scored 837 points. How many more points does Roger need to score? (Lesson 1.7)

   (A) 1,163 points

   (B) 1,173 points

   (C) 1,237 points

   (D) 2,837 points

4. Sue wants to use mental math to find 7×53. Which expression could she use? (Lesson 2.5)

   (A) $(7 \times 5) + 3$

   (B) $(7 \times 5) + (7 \times 3)$

   (C) $(7 \times 50) + 3$

   (D) $(7 \times 50) + (7 \times 3)$

5. Pat listed numbers that all have 15 as a multiple. Which of the following could be Pat's list? (Lesson 5.4)

   (A) 1, 3, 5, 15

   (B) 1, 5, 10, 15

   (C) 1, 15, 30, 45

   (D) 15, 115, 215

6. Which is a true statement about 7 and 14? (Lesson 5.4)

   (A) 7 is a multiple of 14.

   (B) 14 is a factor of 7.

   (C) 14 is a common multiple of 7 and 14.

   (D) 21 is a common multiple of 7 and 14.

Name _____

Chapter 5 Extra Practice

Lesson 5.1

Use tiles to find all the factors of the product. Record the arrays on grid paper and write the factors shown.

1. 17   2. 42   3. 28   4. 50

_____   _____   _____   _____

_____   _____   _____   _____

Lesson 5.2

Is 5 a factor of the number? Write *yes* or *no*.

1. 35   2. 56   3. 51   4. 40

_____   _____   _____   _____

List all the factor pairs in the table.

5.	Factors of 16	
_____ × _____ = _____	_____, _____
_____ × _____ = _____	_____, _____
_____ × _____ = _____	_____, _____

6.	Factors of 49	
_____ × _____ = _____	_____, _____
_____ × _____ = _____	_____, _____

Lesson 5.3

Solve.

1. Hana is putting the fruit she bought into bowls. She bought 8 melons, 12 pears, and 24 apples. She puts the same number of pieces of fruit in each bowl and puts only one type of fruit in each bowl. How many pieces can Hana put in each bowl?

2. A store owner is arranging clothing on racks. She has 30 sweaters, 45 shirts, and 15 pairs of jeans. She wants to put the same number of items on each rack, with only one type of item on each. How many items can she put on a rack?

_____   _____

Lesson 5.4

Is the number a multiple of 9? Write *yes* or *no*.

1. 24   **2.** 18   **3.** 27   **4.** 42

_____   _____

List the next nine multiples of each number.
Find the common multiples.

5. Multiples of 4: 4, _____

Multiples of 5: 5, _____

Common multiples: _____

6. Multiples of 3: 3, _____

Multiples of 6: 6, _____

Common multiples: _____

Lesson 5.5

Tell whether the number is *prime* or *composite*.

1. 39   **2.** 29   **3.** 51

_____   _____   _____

Lesson 5.6

Use the rule to write the first twelve numbers in the pattern.
Describe another pattern in the numbers.

1. Rule: Add 6.   First term: 10

_____

_____

2. Rule: Add 3, subtract 2.   First term: 7

_____

_____

_____

_____

common denominator A common multiple of the denominators of two or more fractions

denominator The part of the fraction below the line, which tells how many equal parts there are in the whole or in a group

equivalent fractions Two or more fractions that name the same amount

numerator The part of a fraction above the line, which tells how many parts are being counted

simplest form A fraction in which 1 is the only number that can divide evenly into the numerator and the denominator

Dear Family,

During the next few weeks, our math class will be learning more about fractions. We will learn how to compare fractions, order fractions, and find equivalent fractions.

You can expect to see homework that provides practice with fractions.

Here is a sample of how your child will be taught to compare fractions that have the same numerator.

🔑 MODEL Compare Fractions with the Same Numerator

This is one way we will be comparing fractions that have the same numerator.

STEP 1

Compare $\frac{4}{10}$ and $\frac{4}{6}$.

Look at the numerators.

Each numerator is 4.

The numerators are the same.

STEP 2

Since the numerators are the same, look at the denominators, 10 and 6.

The more pieces a whole is divided into, the smaller the pieces are. Tenths are smaller pieces than sixths.

So, $\frac{4}{10}$ is a smaller fraction of the whole than $\frac{4}{6}$.

$\frac{4}{10}$ is less than $\frac{4}{6}$. $\frac{4}{10} < \frac{4}{6}$

Tips

Identifying Fewer Pieces

The fewer pieces a whole is divided into, the larger the pieces are. For example, when a whole is divided into 6 equal pieces, the pieces are larger than when the same size whole is divided into 10 equal pieces. So, $\frac{4}{6}$ is greater than (>) $\frac{4}{10}$.

Activity

Play a card game to help your child practice comparing fractions. On several cards, write a pair of fractions with the same numerator and draw a circle between the fractions. Players take turns drawing a card and telling whether *greater than* (>) or *less than* (<) belongs in the circle.

Carta
para la casa

Vocabulario

común denominador Un múltiplo común de dos o más denominadores

denominador La parte de la fracción debajo de la barra que indica cuántas partes iguales hay en un total o en un grupo

fracciones equivalentes Dos o más fracciones que representan la misma cantidad

mínima expresión Una fracción en la que 1 es el único número que se puede dividir en partes iguales entre el numerador y el denominador

numerador La parte de una fracción por encima de la barra que indica cuántas partes se están contando

Querida familia,

Durante las próximas semanas, en la clase de matemáticas aprenderemos más sobre las fracciones. Aprenderemos a comparar y ordenar fracciones, y a hallar fracciones equivalentes.

Llevaré a la casa tareas para practicar las fracciones.

Este es un ejemplo de la manera como aprenderemos a comparar fracciones que tienen el mismo numerador.

🔑 MODELO   Comparar fracciones que tienen el mismo numerador

Esta es una manera como compararemos fracciones que tienen el mismo numerador.

Paso 1

Compara $\frac{4}{10}$ y $\frac{4}{6}$.

Mira los numeradores.

Cada numerador es 4.

Los numeradores son iguales.

Paso 2

Dado que los numeradores son iguales, Mira los denominadores 10 y 6.

Entre más piezas se divida un entero, las piezas serán más pequeñas. Los décimos son piezas más pequeñas que los sextos.

Por lo tanto, $\frac{4}{10}$ es una fracción menor del entero que $\frac{4}{6}$.

$\frac{4}{10}$ es menor que $\frac{4}{6}$. $\frac{4}{10} < \frac{4}{6}$

Pistas

Identificar menos piezas

Entre menos piezas se divida un entero, las piezas serán más grandes. Por ejemplo, si un entero se divide en 6 piezas iguales, las piezas son más grandes que las piezas del mismo entero, si éste se divide en 10 piezas iguales. Por lo tanto, $\frac{4}{6}$ es mayor que (>) $\frac{4}{10}$.

Actividad

Ayude a su hijo a comparar fracciones jugando con tarjetas de fracciones. En varias tarjetas, escriba pares de fracciones con el mismo numerador y dibuje un círculo entre las fracciones. Túrnense para dibujar cada tarjeta y decir qué debe ir en el círculo: "mayor que" o "menor que."

Equivalent Fractions

Use the model to write an equivalent fraction.

1.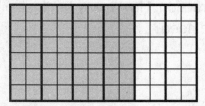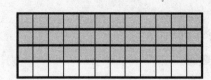

$$\frac{4}{6}$$ = $$\frac{2}{3}$$ _____

2.

$$\frac{3}{4}$$ = _____

Tell whether the fractions are equivalent. Write = or ≠.

3. $\frac{8}{10} \bigcirc \frac{4}{5}$ 4. $\frac{1}{2} \bigcirc \frac{7}{12}$ 5. $\frac{3}{4} \bigcirc \frac{8}{12}$ 6. $\frac{2}{3} \bigcirc \frac{4}{6}$

7. $\frac{5}{8} \bigcirc \frac{4}{10}$ 8. $\frac{2}{6} \bigcirc \frac{4}{12}$ 9. $\frac{20}{100} \bigcirc \frac{1}{5}$ 10. $\frac{5}{8} \bigcirc \frac{9}{10}$

Problem Solving  REAL WORLD

11. Jamal finished $\frac{5}{6}$ of his homework. Margaret finished $\frac{3}{4}$ of her homework, and Steve finished $\frac{10}{12}$ of his homework. Which two students finished the same amount of homework?

12. Sophia's vegetable garden is divided into 12 equal sections. She plants carrots in 8 of the sections. Write two fractions that are equivalent to the part of Sophia's garden that is planted with carrots.

_____

_____

Lesson Check

1. A rectangle is divided into 8 equal parts. Two parts are shaded. Which fraction is equivalent to the shaded area of the rectangle?

   (A) $\frac{1}{4}$

   (B) $\frac{1}{3}$

   (C) $\frac{2}{6}$

   (D) $\frac{3}{4}$

2. Jeff uses 3 fifth-size strips to model $\frac{3}{5}$. He wants to use tenth-size strips to model an equivalent fraction. How many tenth-size strips will he need?

   (A) 10

   (B) 6

   (C) 5

   (D) 3

Spiral Review

3. Cassidy places 40 stamps on each of 8 album pages. How many stamps does she place in all? **(Lesson 2.3)**

   (A) 300

   (B) 320

   (C) 360

   (D) 380

4. Maria and 3 friends have 1,200 soccer cards. If they share the soccer cards equally, how many will each person receive? **(Lesson 4.4)**

   (A) 30   (C) 300

   (B) 40   (D) 400

5. Six groups of students sell 162 balloons at the school carnival. There are 3 students in each group. If each student sells the same number of balloons, how many balloons does each student sell?

   (Lesson 4.12)

   (A) 9

   (B) 18

   (C) 27

   (D) 54

6. Four students each made a list of prime numbers.

   Eric: 5, 7, 17, 23

   Maya: 3, 5, 13, 17

   Bella: 2, 3, 17, 19

   Jordan: 7, 11, 13, 21

   Who made an error and included a composite number? **(Lesson 5.5)**

   (A) Eric

   (B) Maya

   (C) Bella

   (D) Jordan

Name _____

Generate Equivalent Fractions

Write two equivalent fractions for each.

1. $\frac{1}{3}$

$\frac{1 \times 2}{3 \times 2} = \frac{2}{6}$

$\frac{1 \times 4}{3 \times 4} = \frac{4}{12}$

2. $\frac{2}{3}$

_____

3. $\frac{1}{2}$

_____

4. $\frac{4}{5}$

_____

Tell whether the fractions are equivalent.
Write = or ≠.

5. $\frac{1}{4} \bigcirc \frac{3}{12}$

6. $\frac{4}{5} \bigcirc \frac{5}{10}$

7. $\frac{3}{8} \bigcirc \frac{2}{6}$

8. $\frac{3}{4} \bigcirc \frac{6}{8}$

9. $\frac{5}{6} \bigcirc \frac{10}{12}$

10. $\frac{6}{12} \bigcirc \frac{5}{8}$

11. $\frac{2}{5} \bigcirc \frac{4}{10}$

12. $\frac{2}{4} \bigcirc \frac{3}{12}$

Problem Solving  REAL WORLD

13. Jan has a 12-ounce milkshake. Four ounces in the milkshake are vanilla, and the rest is chocolate. What are two equivalent fractions that represent the fraction of the milkshake that is vanilla?

_____

14. Kareem lives $\frac{4}{10}$ of a mile from the mall. Write two equivalent fractions that show what fraction of a mile Kareem lives from the mall.

_____

Lesson Check

1. Jessie colored a poster. She colored $\frac{2}{5}$ of the poster red. Which fraction is equivalent to $\frac{2}{5}$?

   (A) $\frac{4}{10}$ (C) $\frac{4}{5}$

   (B) $\frac{7}{10}$ (D) $\frac{2}{2}$

2. Marcus makes a punch that is $\frac{1}{4}$ cranberry juice. Which two fractions are equivalent to $\frac{1}{4}$?

   (A) $\frac{2}{5}, \frac{3}{12}$ (C) $\frac{3}{4}, \frac{6}{8}$

   (B) $\frac{2}{8}, \frac{4}{12}$ (D) $\frac{2}{8}, \frac{3}{12}$

Spiral Review

3. An electronics store sells a large flat screen television for $1,699. Last month, the store sold 8 of these television sets. About how much money did the store make on the television sets? (Lesson 2.4)

   (A) $160,000

   (B) $16,000

   (C) $8,000

   (D) $1,600

4. Matthew has 18 sets of baseball cards. Each set has 12 cards. About how many baseball cards does Matthew have in all? (Lesson 3.2)

   (A) 300

   (B) 200

   (C) 150

   (D) 100

5. Diana had 41 stickers. She put them in 7 equal groups. She put as many as possible in each group. She gave the leftover stickers to her sister. How many stickers did Diana give to her sister? (Lesson 4.3)

   (A) 3

   (B) 4

   (C) 5

   (D) 6

6. Christopher wrote the number pattern below. The first term is 8.
   8, 6, 9, 7, 10, …
   Which is a rule for the pattern? (Lesson 5.6)

   (A) Add 2, add 3.

   (B) Add 6, subtract 3.

   (C) Subtract 6, add 3.

   (D) Subtract 2, add 3.

Simplest Form

Write the fraction in simplest form.

1. $\dfrac{6}{10}$

2. $\dfrac{6}{8}$

3. $\dfrac{5}{5}$

4. $\dfrac{8}{12}$

$$\dfrac{6}{10} = \dfrac{6 \div 2}{10 \div 2} = \dfrac{3}{5}$$

_____ _____ _____

5. $\dfrac{100}{100}$

6. $\dfrac{2}{6}$

7. $\dfrac{2}{8}$

8. $\dfrac{4}{10}$

_____ _____ _____ _____

Tell whether the fractions are equivalent.
Write = or ≠.

9. $\dfrac{6}{12} \bigcirc \dfrac{1}{12}$

10. $\dfrac{3}{4} \bigcirc \dfrac{5}{6}$

11. $\dfrac{6}{10} \bigcirc \dfrac{3}{5}$

12. $\dfrac{3}{12} \bigcirc \dfrac{1}{3}$

13. $\dfrac{6}{10} \bigcirc \dfrac{60}{100}$

14. $\dfrac{11}{12} \bigcirc \dfrac{9}{10}$

15. $\dfrac{2}{5} \bigcirc \dfrac{8}{20}$

16. $\dfrac{4}{8} \bigcirc \dfrac{1}{2}$

Problem Solving REAL WORLD

17. At Memorial Hospital, 9 of the 12 babies born on Tuesday were boys. In simplest form, what fraction of the babies born on Tuesday were boys?

18. Cristina uses a ruler to measure the length of her math textbook. She says that the book is $\dfrac{4}{10}$ meter long. Is her measurement in simplest form? If not, what is the length of the book in simplest form?

_____   _____

Lesson Check

1. Six out of the 12 members of the school choir are boys. In simplest form, what fraction of the choir is boys?

    (A) $\frac{1}{6}$

    (B) $\frac{6}{12}$

    (C) $\frac{1}{2}$

    (D) $\frac{12}{6}$

2. Which of the following fractions is in simplest form?

    (A) $\frac{5}{6}$

    (B) $\frac{6}{8}$

    (C) $\frac{8}{10}$

    (D) $\frac{2}{12}$

Spiral Review

3. Each of the 23 students in Ms. Evans' class raised $45 for the school by selling coupon books. How much money did the class raise in all? **(Lesson 3.5)**

    (A) $207   (C) $1,025

    (B) $225   (D) $1,035

4. Which pair of numbers below have 4 and 6 as common factors? **(Lesson 5.3)**

    (A) 12, 18

    (B) 20, 24

    (C) 28, 30

    (D) 36, 48

5. Bart uses $\frac{3}{12}$ cup milk to make muffins. Which fraction is equivalent to $\frac{3}{12}$?

    (Lesson 6.2)

    (A) $\frac{1}{4}$

    (B) $\frac{1}{3}$

    (C) $\frac{1}{2}$

    (D) $\frac{2}{3}$

6. Ashley bought 4 packages of juice boxes. There are 6 juice boxes in each package. She gave 2 juice boxes to each of 3 friends. How many juice boxes does Ashley have left? **(Lesson 2.12)**

    (A) 24

    (B) 22

    (C) 18

    (D) 12

Common Denominators

Write the pair of fractions as a pair of fractions with a common denominator.

1. $\frac{2}{3}$ and $\frac{3}{4}$

Think: Find a common multiple.
3: 3, 6, 9, ⑫, 15
4: 4, 8, ⑫, 16, 20
$$\frac{8}{12}, \frac{9}{12}$$

2. $\frac{1}{4}$ and $\frac{2}{3}$

_____

3. $\frac{3}{10}$ and $\frac{1}{2}$

_____

4. $\frac{3}{5}$ and $\frac{3}{4}$

5. $\frac{2}{4}$ and $\frac{7}{8}$

6. $\frac{2}{3}$ and $\frac{5}{12}$

7. $\frac{1}{4}$ and $\frac{1}{6}$

_____ _____ _____ _____

Tell whether the fractions are equivalent. Write = or ≠.

8. $\frac{1}{2} \bigcirc \frac{2}{5}$

9. $\frac{1}{2} \bigcirc \frac{3}{6}$

10. $\frac{3}{4} \bigcirc \frac{5}{6}$

11. $\frac{6}{10} \bigcirc \frac{3}{5}$

12. $\frac{6}{8} \bigcirc \frac{3}{4}$

13. $\frac{3}{4} \bigcirc \frac{2}{3}$

14. $\frac{2}{10} \bigcirc \frac{4}{5}$

15. $\frac{1}{4} \bigcirc \frac{3}{12}$

Problem Solving REAL WORLD

16. Adam drew two same size rectangles and divided them into the same number of equal parts. He shaded $\frac{1}{3}$ of one rectangle and $\frac{1}{4}$ of other rectangle. What is the least number of parts into which both rectangles could be divided?

17. Mera painted equal sections of her bedroom wall to make a pattern. She painted $\frac{2}{5}$ of the wall white and $\frac{1}{2}$ of the wall lavender. Write an equivalent fraction for each using a common denominator.

_____ _____

Lesson Check

1. Which of the following is a common denominator of $\frac{1}{4}$ and $\frac{5}{6}$?

    (A) 8

    (B) 9

    (C) 12

    (D) 15

2. Two fractions have a common denominator of 8. Which of the following could be the two fractions?

    (A) $\frac{1}{2}$ and $\frac{2}{3}$

    (B) $\frac{1}{4}$ and $\frac{1}{2}$

    (C) $\frac{3}{4}$ and $\frac{1}{6}$

    (D) $\frac{1}{2}$ and $\frac{4}{5}$

Spiral Review

3. Which number is 100,000 more than seven hundred two thousand, eighty-three? (Lesson 1.2)

    (A) 703,083

    (B) 712,083

    (C) 730,083

    (D) 802,083

4. Aiden baked 8 dozen muffins. How many total muffins did he bake? (Lesson 2.10)

    (A) 64

    (B) 80

    (C) 96

    (D) 104

5. On a bulletin board, the principal, Ms. Gomez, put 115 photos of the fourth-grade students in her school. She put the photos in 5 equal rows. How many photos did she put in each row? (Lesson 4.11)

    (A) 21

    (B) 23

    (C) 25

    (D) 32

6. Judy uses 12 tiles to make a mosaic. Eight of the tiles are blue. What fraction, in simplest form, represents the tiles that are blue? (Lesson 6.3)

    (A) $\frac{2}{3}$

    (B) $\frac{2}{5}$

    (C) $\frac{3}{4}$

    (D) $\frac{12}{18}$

Name _____

Problem Solving • Find Equivalent Fractions

Solve each problem.

1. Miranda is braiding her hair. Then she will attach beads to the braid. She wants $\frac{1}{3}$ of the beads to be red. If the greatest number of beads that will fit on the braid is 12, what other fractions could represent the part of the beads that are red?

$$\frac{2}{6}, \frac{3}{9}, \frac{4}{12}$$

2. Ms. Groves has trays of paints for students in her art class. Each tray has 5 colors. One of the colors is purple. What fraction of the colors in 20 trays is purple?

_____

3. Miguel is making an obstacle course for field day. At the end of every sixth of the course, there is a tire. At the end of every third of the course, there is a cone. At the end of every half of the course, there is a hurdle. At which locations of the course will people need to go through more than one obstacle?

_____

4. Preston works in a bakery where he puts muffins in boxes. He makes the following table to remind himself how many blueberry muffins should go in each box.

Number of Blueberry Muffins	2	4	8	■
Total Number of Muffins	6	12	24	36

How many blueberry muffins should Preston put in a box with 36 muffins?

_____

Lesson Check

1. A used bookstore will trade 2 of its books for 3 of yours. If Val brings in 18 books to trade, how many books can she get from the store?

   (A) 9

   (B) 12

   (C) 18

   (D) 27

2. Every $\frac{1}{2}$ hour Naomi stretches her neck; every $\frac{1}{3}$ hour she stretches her legs; and every $\frac{1}{6}$ hour she stretches her arms. Which parts of her body will Naomi stretch when $\frac{2}{3}$ of an hour has passed?

   (A) neck and legs

   (B) neck and arms

   (C) legs and arms

   (D) none

Spiral Review

3. At the beginning of the year, the Wong family car had been driven 14,539 miles. At the end of the year, their car had been driven 21,844 miles. How many miles did the Wong family drive their car during that year? (Lesson 1.7)

   (A) 6,315 miles

   (B) 7,295 miles

   (C) 7,305 miles

   (D) 36,383 miles

4. Widget Company made 3,600 widgets in 4 hours. They made the same number of widgets each hour. How many widgets did the company make in one hour?
   (Lesson 4.4)

   (A) 80

   (B) 90

   (C) 800

   (D) 900

5. Tyler is thinking of a number that is divisible by 2 and by 3. By which of the following numbers must Tyler's number also be divisible? (Lesson 5.2)

   (A) 6

   (B) 8

   (C) 9

   (D) 12

6. Jessica drew a circle divided into 8 equal parts. She shaded 6 of the parts. Which fraction is equivalent to the part of the circle that is shaded? (Lesson 6.1)

   (A) $\frac{2}{3}$

   (B) $\frac{3}{4}$

   (C) $\frac{10}{16}$

   (D) $\frac{12}{18}$

Compare Fractions Using Benchmarks

Compare. Write < or >.

1. $\frac{1}{8}$ ⟨<⟩ $\frac{6}{10}$

Think: $\frac{1}{8}$ is less than $\frac{1}{2}$.

$\frac{6}{10}$ is more than $\frac{1}{2}$.

2. $\frac{4}{12}$ ◯ $\frac{4}{6}$

3. $\frac{2}{8}$ ◯ $\frac{1}{2}$

4. $\frac{3}{5}$ ◯ $\frac{3}{3}$

5. $\frac{7}{8}$ ◯ $\frac{5}{10}$

6. $\frac{9}{12}$ ◯ $\frac{1}{3}$

7. $\frac{4}{6}$ ◯ $\frac{7}{8}$

8. $\frac{2}{4}$ ◯ $\frac{2}{3}$

9. $\frac{3}{5}$ ◯ $\frac{1}{4}$

10. $\frac{6}{10}$ ◯ $\frac{2}{5}$

11. $\frac{1}{8}$ ◯ $\frac{2}{10}$

12. $\frac{2}{3}$ ◯ $\frac{5}{12}$

13. $\frac{4}{5}$ ◯ $\frac{5}{6}$

14. $\frac{3}{5}$ ◯ $\frac{5}{8}$

15. $\frac{8}{8}$ ◯ $\frac{3}{4}$

Problem Solving  REAL WORLD

16. Erika ran $\frac{3}{8}$ mile. Maria ran $\frac{3}{4}$ mile. Who ran farther?

_____

17. Carlos finished $\frac{1}{3}$ of his art project on Monday. Tyler finished $\frac{1}{2}$ of his art project on Monday. Who finished more of his art project on Monday?

_____

Lesson Check

1. Which symbol makes the statement true?

   $$\frac{4}{6} \bullet \frac{3}{8}$$

   (A) >

   (B) <

   (C) =

   (D) none

2. Which of the following fractions is greater than $\frac{3}{4}$?

   (A) $\frac{1}{4}$

   (B) $\frac{5}{6}$

   (C) $\frac{3}{8}$

   (D) $\frac{2}{3}$

Spiral Review

3. Abigail is putting tiles on a table top. She needs 48 tiles for each of 8 rows. Each row will have 6 white tiles. The rest of the tiles will be purple. How many purple tiles will she need? **(Lesson 2.9)**

   (A) 432

   (B) 384

   (C) 336

   (D) 48

4. Each school bus going on the field trip holds 36 students and 4 adults. There are 6 filled buses on the field trip. How many people are going on the field trip? **(Lesson 2.9)**

   (A) 216

   (B) 240

   (C) 256

   (D) 360

5. Noah wants to display his 72 collector's flags. He is going to put 6 flags in each row. How many rows of flags will he have in his display? **(Lesson 4.7)**

   (A) 12

   (B) 15

   (C) 18

   (D) 21

6. Julian wrote this number pattern on the board:

   3, 10, 17, 24, 31, 38.

   Which of the numbers in Julian's pattern are composite numbers? **(Lesson 5.5)**

   (A) 3, 17, 31

   (B) 10, 24, 38

   (C) 10, 17, 38

   (D) 17, 24, 38

Compare Fractions

Compare. Write <, >, or =.

1. $\dfrac{3}{4}$ $<$ $\dfrac{5}{6}$

Think: 12 is a common denominator.

$$\dfrac{3}{4} = \dfrac{3 \times 3}{4 \times 3} = \dfrac{9}{12}$$

$$\dfrac{5}{6} = \dfrac{5 \times 2}{6 \times 2} = \dfrac{10}{12}$$

$$\dfrac{9}{12} < \dfrac{10}{12}$$

2. $\dfrac{1}{5}$ ◯ $\dfrac{2}{10}$

3. $\dfrac{2}{4}$ ◯ $\dfrac{2}{5}$

4. $\dfrac{3}{5}$ ◯ $\dfrac{7}{10}$

5. $\dfrac{4}{12}$ ◯ $\dfrac{1}{6}$

6. $\dfrac{2}{6}$ ◯ $\dfrac{1}{3}$

7. $\dfrac{1}{3}$ ◯ $\dfrac{2}{4}$

8. $\dfrac{2}{5}$ ◯ $\dfrac{1}{2}$

9. $\dfrac{4}{8}$ ◯ $\dfrac{2}{4}$

10. $\dfrac{7}{12}$ ◯ $\dfrac{2}{4}$

11. $\dfrac{1}{8}$ ◯ $\dfrac{3}{4}$

Problem Solving REAL WORLD

12. A recipe uses $\dfrac{2}{3}$ cup of flour and $\dfrac{5}{8}$ cup of blueberries. Is there more flour or more blueberries in the recipe?

13. Peggy completed $\dfrac{5}{6}$ of the math homework and Al completed $\dfrac{4}{5}$ of the math homework. Did Peggy or Al complete more of the math homework?

_____

_____

Lesson Check

1. Pedro fills a glass $\frac{2}{4}$ full with orange juice. Which of the following fractions is greater than $\frac{2}{4}$?

   (A) $\frac{3}{8}$

   (B) $\frac{4}{6}$

   (C) $\frac{5}{12}$

   (D) $\frac{1}{3}$

2. Today Ian wants to run less than $\frac{7}{12}$ mile. Which of the following distances is less than $\frac{7}{12}$ mile?

   (A) $\frac{3}{4}$ mile

   (B) $\frac{2}{3}$ mile

   (C) $\frac{5}{6}$ mile

   (D) $\frac{2}{4}$ mile

Spiral Review

3. Ms. Davis traveled 372,645 miles last year on business. What is the value of 6 in 372,645? (Lesson 1.1)

   (A)   6

   (B)   60

   (C)   600

   (D) 6,000

4. One section of an auditorium has 12 rows of seats. Each row has 13 seats. What is the total number of seats in that section? (Lesson 3.4)

   (A) 25

   (B) 144

   (C) 156

   (D) 169

5. Sam has 12 black-and-white photos and 18 color photos. He wants to put the photos in equal rows so each row has either black-and-white photos only or color photos only. In how many rows can Sam arrange the photos? (Lesson 5.3)

   (A) 1, 2, 3, or 6 rows

   (B) 1, 3, 6, or 9 rows

   (C) 1, 2, or 4 rows

   (D) 1, 2, 3, 4, 6, or 9 rows

6. The teacher writes $\frac{10}{12}$ on the board. He asks students to write the fraction in simplest form. Who writes the correct answer? (Lesson 6.3)

   (A) JoAnn writes $\frac{10}{12}$.

   (B) Karen writes $\frac{5}{12}$.

   (C) Lynn writes $\frac{6}{5}$.

   (D) Mark writes $\frac{5}{6}$.

Compare and Order Fractions

Write the fractions in order from least to greatest.

1. $\frac{5}{8}, \frac{2}{12}, \frac{8}{10}$

   Use benchmarks and a number line.

   Think: $\frac{5}{8}$ is close to $\frac{1}{2}$. $\frac{2}{12}$ is close to 0.

   $\frac{8}{10}$ is close to 1.

   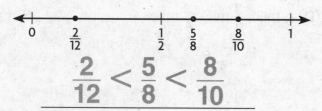

   $$\frac{2}{12} < \frac{5}{8} < \frac{8}{10}$$

2. $\frac{1}{5}, \frac{2}{3}, \frac{5}{8}$

   _____

3. $\frac{1}{2}, \frac{2}{5}, \frac{6}{10}$

   _____

4. $\frac{4}{6}, \frac{7}{12}, \frac{5}{10}$

   _____

5. $\frac{1}{4}, \frac{3}{6}, \frac{1}{8}$

   _____

6. $\frac{1}{8}, \frac{3}{6}, \frac{7}{12}$

   _____

7. $\frac{8}{100}, \frac{3}{5}, \frac{7}{10}$

   _____

8. $\frac{3}{4}, \frac{7}{8}, \frac{1}{5}$

   _____

Problem Solving REAL WORLD

9. Amy's math notebook weighs $\frac{1}{2}$ pound, her science notebook weighs $\frac{7}{8}$ pound, and her history notebook weighs $\frac{3}{4}$ pound. What are the weights in order from lightest to heaviest?

   _____

   _____

10. Carl has three picture frames. The thicknesses of the frames are $\frac{4}{5}$ inch, $\frac{3}{12}$ inch, and $\frac{5}{6}$ inch. What are the thicknesses in order from least to greatest?

    _____

    _____

Lesson Check

1. Juan's three math quizzes this week took him $\frac{1}{3}$ hour, $\frac{4}{6}$ hour, and $\frac{1}{5}$ hour to complete. Which list shows the lengths of time in order from least to greatest?

   (A) $\frac{1}{3}$ hour, $\frac{4}{6}$ hour, $\frac{1}{5}$ hour

   (B) $\frac{1}{5}$ hour, $\frac{1}{3}$ hour, $\frac{4}{6}$ hour

   (C) $\frac{1}{3}$ hour, $\frac{1}{5}$ hour, $\frac{4}{6}$ hour

   (D) $\frac{4}{6}$ hour, $\frac{1}{3}$ hour, $\frac{1}{5}$ hour

2. On three days last week, Maria ran $\frac{3}{4}$ mile, $\frac{7}{8}$ mile, and $\frac{3}{5}$ mile. What are the distances in order from least to greatest?

   (A) $\frac{3}{4}$ mile, $\frac{7}{8}$ mile, $\frac{3}{5}$ mile

   (B) $\frac{3}{5}$ mile, $\frac{3}{4}$ mile, $\frac{7}{8}$ mile

   (C) $\frac{7}{8}$ mile, $\frac{3}{4}$ mile, $\frac{3}{5}$ mile

   (D) $\frac{7}{8}$ mile, $\frac{3}{5}$ mile, $\frac{3}{4}$ mile

Spiral Review

3. Santiago collects 435 cents in nickels. How many nickels does he collect? (Lesson 4.5)

   (A) 58

   (B) 78

   (C) 85

   (D) 87

4. Lisa has three classes that each last 50 minutes. What is the total number of minutes the three classes last? (Lesson 3.1)

   (A) 15 minutes

   (B) 150 minutes

   (C) 153 minutes

   (D) 156 minutes

5. Some students were asked to write a composite number. Which student did NOT write a composite number? (Lesson 5.5)

   (A) Alicia wrote 2.

   (B) Bob wrote 9.

   (C) Arianna wrote 15.

   (D) Daniel wrote 21.

6. Mrs. Carmel serves $\frac{6}{8}$ of a loaf of bread with dinner. Which fraction is equivalent to $\frac{6}{8}$? (Lesson 6.2)

   (A) $\frac{2}{4}$

   (B) $\frac{9}{16}$

   (C) $\frac{2}{3}$

   (D) $\frac{3}{4}$

Chapter 6 Extra Practice

Lesson 6.1

Tell whether the fractions are equivalent. Write = or ≠.

1. $\frac{5}{10}$ ◯ $\frac{1}{2}$
2. $\frac{2}{3}$ ◯ $\frac{3}{6}$
3. $\frac{6}{8}$ ◯ $\frac{3}{4}$
4. $\frac{7}{12}$ ◯ $\frac{4}{6}$

Lesson 6.2

Write two equivalent fractions for each.

1. $\frac{2}{3}$
2. $\frac{5}{10}$
3. $\frac{4}{12}$
4. $\frac{4}{5}$

_____   _____   _____

Lesson 6.3

Write the fraction in simplest form.

1. $\frac{6}{12}$
2. $\frac{2}{10}$
3. $\frac{4}{6}$
4. $\frac{3}{12}$
5. $\frac{6}{10}$

_____   _____   _____   _____   _____

Lesson 6.4

Write the pair of fractions as a pair of fractions with a common denominator.

1. $\frac{2}{3}$ and $\frac{5}{6}$
2. $\frac{3}{5}$ and $\frac{1}{2}$
3. $\frac{1}{4}$ and $\frac{5}{12}$

_____   _____   _____

4. $\frac{7}{8}$ and $\frac{3}{4}$
5. $\frac{3}{10}$ and $\frac{1}{5}$
6. $\frac{3}{4}$ and $\frac{1}{3}$

_____   _____   _____

Lesson 6.5

1. Mr. Renner is decorating a bulletin board with groups of shapes. Each group has 3 shapes, and $\frac{2}{3}$ of the shapes are snowflakes. If Mr. Renner is using 4 groups of shapes, how many snowflakes will he need?

Complete the table to find the fraction of the shapes for each number of group that are snowflakes.

Groups of Shapes	1	2	3	
Number of Snowflakes / Number of Shapes	$\frac{2}{3}$	$\frac{4}{\square}$		

How many snowflake shapes will Mr. Renner use? _____

2. Nell made a pizza. She cut the pizza into fourths. Then she cut each fourth into four pieces. Nell and her friends ate 6 of the smaller pieces of the pizza.

What fraction of the pizza did Nell and her friends eat? _____

What fraction of the pizza did Nell and her friends NOT eat? _____

Lessons 6.6 - 6.7

Compare. Write <, >, or =.

1. $\frac{2}{6} \bigcirc \frac{3}{4}$

2. $\frac{6}{8} \bigcirc \frac{1}{4}$

3. $\frac{5}{6} \bigcirc \frac{2}{4}$

4. $\frac{1}{3} \bigcirc \frac{4}{12}$

5. $\frac{1}{6} \bigcirc \frac{1}{8}$

6. $\frac{2}{3} \bigcirc \frac{4}{6}$

7. $\frac{3}{10} \bigcirc \frac{3}{12}$

8. $\frac{7}{8} \bigcirc \frac{4}{4}$

Lesson 6.8

Write the fractions in order from least to greatest.

1. $\frac{1}{2}, \frac{1}{4}, \frac{5}{8}$

2. $\frac{2}{3}, \frac{1}{6}, \frac{9}{10}$

3. $\frac{3}{5}, \frac{3}{4}, \frac{3}{8}$

_____   _____   _____

Vocabulary

denominator The number in a fraction that tells how many equal parts are in the whole or in the group

fraction A number that names a part of a whole or part of a group

mixed number A number represented by a whole number and a fraction

numerator The number in a fraction that tells how many parts of the whole or group are being considered

unit fraction A fraction that has a numerator of 1

Dear Family,

During the next few weeks, our math class will be learning how to add and subtract fractions and mixed numbers. First, we will use models to find the sums or the differences. Then we will record equations to match our models. Finally, we will add and subtract without using models.

You can expect to see homework that provides practice adding and subtracting fractions with and without models.

Here is a sample of how your child will be taught to add fractions using fraction strips.

🔒 MODEL Add Fractions Using Models

This is how we will be adding fractions using fraction strips.

Model $\frac{1}{6} + \frac{3}{6}$.

1

$\frac{1}{6}$	$\frac{1}{6}$	$\frac{1}{6}$	$\frac{1}{6}$	$\frac{1}{6}$	$\frac{1}{6}$

$\underbrace{}_{\frac{1}{6}}$ $\underbrace{}_{\frac{3}{6}}$

STEP 1

Each section represents 1 sixth. How many sixths are there in all?
4 sixths

STEP 2

Write the number of sixths as a fraction.

4 sixths = $\frac{4}{6}$

$\frac{1}{6} + \frac{3}{6} = \frac{4}{6}$

Tips

Renaming as a Mixed Number

When the numerator is greater than the denominator, you can rename the sum or the difference as a mixed number.

$$\frac{9}{8} = \frac{8}{8} + \frac{1}{8}$$
$$= 1 + \frac{1}{8}$$
$$= 1\frac{1}{8}$$

Activity

Have your child use measuring cups to practice addition and subtraction of fractions. For example, to model $\frac{1}{4} + \frac{3}{4}$, have your child use rice to fill one measuring cup to the $\frac{1}{4}$-cup mark and another measuring cup to the $\frac{3}{4}$-cup mark. Then ask him or her to combine the amounts to find the sum, $\frac{4}{4}$ or 1 whole cup.

© Houghton Mifflin Harcourt Publishing Company

Carta para la casa

Vocabulario

denominador El número de una fracción que dice cuántas partes iguales hay en el todo o en el grupo

fracción Un número que nombra una parte de un todo o una parte de un grupo

número mixto Un número representado por un número entero y una fracción

numerador El número de una fracción que dice cuántas partes del todo o de un grupo están siendo consideradas

fracción unitaria Una fracción cuyo numerador es 1

Querida familia,

Durante las próximas semanas, en la clase de matemáticas estudiaremos la suma y resta de fracciones y números mixtos. Primero usaremos modelos para hallar las sumas o las diferencias. Después haremos ecuaciones que se ajusten a nuestros modelos. Finalmente, sumaremos y restaremos sin usar modelos.

Llevaré a casa tareas con actividades para practicar la suma y la resta de fracciones con y sin modelos.

Este es un ejemplo de la manera como aprenderemos a sumar fracciones usando tiras de fracciones.

🔑 MODELO Sumar fracciones usando modelos

Así sumaremos fracciones usando tiras de fracciones.

Representa $\frac{1}{6} + \frac{3}{6}$.

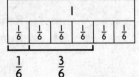

PASO 1

Cada sección representa 1 sexto. ¿Cuántos sextos hay en total?

4 sextos

PASO 2

Escribe el número de sextos como una fracción.

$4 \text{ sextos} = \frac{4}{6}$

$\frac{1}{6} + \frac{3}{6} = \frac{4}{6}$

Pistas

Expresar como un número mixto

Cuando el numerador es mayor que el denominador, puedes expresar la suma o la diferencia como un número mixto.

$$\frac{9}{8} = \frac{8}{8} + \frac{1}{8}$$
$$= 1 + \frac{1}{8}$$
$$= 1\frac{1}{8}$$

Actividad

Pida a su hijo/a que use tazas de medir para practicar la suma y la resta de fracciones. Por ejemplo, para hacer un modelo de $\frac{1}{4} + \frac{3}{4}$, pida a su hijo/a que use arroz para llenar una taza de medir hasta la marca de $\frac{1}{4}$ y otra hasta la marca de $\frac{3}{4}$. Luego pídale que combine las cantidades para hallar la suma, $\frac{4}{4}$ o 1 taza completa.

Add and Subtract Parts of a Whole

Use the model to write an equation.

1.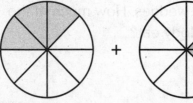

Think: $\dfrac{3}{8}$ $+$ $\dfrac{2}{8}$ $=$ $\dfrac{5}{8}$

$\dfrac{3}{8} + \dfrac{2}{8} = \dfrac{5}{8}$

2.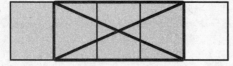

_____

3.

_____

Use the model to solve the equation.

4.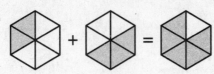

$\dfrac{2}{6} + \dfrac{3}{6} =$ _____

5.

$\dfrac{3}{5} - \dfrac{2}{5} =$ _____

Problem Solving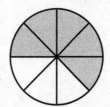

6. Jake ate $\dfrac{4}{8}$ of a pizza. Millie ate $\dfrac{3}{8}$ of the same pizza. How much of the pizza was eaten by Jake and Millie?

_____

_____

7. Kate ate $\dfrac{1}{4}$ of her orange. Ben ate $\dfrac{2}{4}$ of his banana. Did Kate and Ben eat $\dfrac{1}{4} + \dfrac{2}{4} = \dfrac{3}{4}$ of their fruit? **Explain.**

_____

_____

_____

Lesson Check

1. A whole pie is cut into 8 equal slices. Three of the slices are served. How much of the pie is left?

    (A) $\frac{1}{8}$

    (B) $\frac{3}{8}$

    (C) $\frac{5}{8}$

    (D) $\frac{7}{8}$

2. An orange is divided into 6 equal wedges. Jody eats 1 wedge. Then she eats 3 more wedges. How much of the orange did Jody eat?

    (A) $\frac{1}{6}$

    (B) $\frac{4}{6}$

    (C) $\frac{5}{6}$

    (D) $\frac{6}{6}$

Spiral Review

3. Which list of distances is in order from least to greatest? (Lesson 6.8)

    (A) $\frac{1}{8}$ mile, $\frac{3}{16}$ mile, $\frac{3}{4}$ mile

    (B) $\frac{3}{4}$ mile, $\frac{1}{8}$ mile, $\frac{3}{16}$ mile

    (C) $\frac{1}{8}$ mile, $\frac{3}{4}$ mile, $\frac{3}{16}$ mile

    (D) $\frac{3}{16}$ mile, $\frac{1}{8}$ mile, $\frac{3}{4}$ mile

4. Jeremy walked $\frac{6}{8}$ of the way to school and ran the rest of the way. What fraction, in simplest form, shows the part of the way that Jeremy walked? (Lesson 6.3)

    (A) $\frac{1}{4}$

    (B) $\frac{3}{8}$

    (C) $\frac{1}{2}$

    (D) $\frac{3}{4}$

5. An elevator starts on the 100th floor of a building. It descends 4 floors every 10 seconds. At what floor will the elevator be 60 seconds after it starts? (Lesson 5.6)

    (A) 60th floor

    (B) 66th floor

    (C) 72nd floor

    (D) 76th floor

6. For a school play, the teacher asked the class to set up chairs in 20 rows with 25 chairs in each row. After setting up all the chairs, they were 5 chairs short. How many chairs did the class set up? (Lesson 3.7)

    (A) 400

    (B) 450

    (C) 495

    (D) 500

Name _____

Write Fractions as Sums

Write the fraction as a sum of unit fractions.

1. $\dfrac{4}{5}$ = $\underline{\dfrac{1}{5} + \dfrac{1}{5} + \dfrac{1}{5} + \dfrac{1}{5}}$

   Think: Add $\frac{1}{5}$ four times.

2. $\dfrac{3}{8}$ = _____

3. $\dfrac{6}{12}$ = _____

4. $\dfrac{4}{4}$ = _____

Write the fraction as a sum of fractions three different ways.

5. $\dfrac{7}{10}$

_____

_____

_____

6. $\dfrac{6}{6}$

_____

_____

_____

Problem Solving REAL WORLD

7. Miguel's teacher asks him to color $\frac{4}{8}$ of his grid. He must use 3 colors: red, blue, and green. There must be more green sections than red sections. How can Miguel color the sections of his grid to follow all the rules?

_____

_____

_____

8. Petra is asked to color $\frac{6}{6}$ of her grid. She must use 3 colors: blue, red, and pink. There must be more blue sections than red sections or pink sections. What are the different ways Petra can color the sections of her grid and follow all the rules?

_____

_____

_____

_____

Lesson Check

1. Jorge wants to write $\frac{4}{5}$ as a sum of unit fractions. Which of the following should he write?

   Ⓐ $\frac{3}{5} + \frac{1}{5}$

   Ⓑ $\frac{2}{5} + \frac{2}{5}$

   Ⓒ $\frac{1}{5} + \frac{1}{5} + \frac{2}{5}$

   Ⓓ $\frac{1}{5} + \frac{1}{5} + \frac{1}{5} + \frac{1}{5}$

2. Which expression is equivalent to $\frac{7}{8}$?

   Ⓐ $\frac{5}{8} + \frac{2}{8} + \frac{1}{8}$

   Ⓑ $\frac{3}{8} + \frac{3}{8} + \frac{1}{8} + \frac{1}{8}$

   Ⓒ $\frac{4}{8} + \frac{2}{8} + \frac{1}{8}$

   Ⓓ $\frac{4}{8} + \frac{2}{8} + \frac{2}{8}$

Spiral Review

3. An apple is cut into 6 equal slices. Nancy eats 2 of the slices. What fraction of the apple is left? (Lesson 7.1)

   Ⓐ $\frac{1}{6}$

   Ⓑ $\frac{2}{6}$

   Ⓒ $\frac{3}{6}$

   Ⓓ $\frac{4}{6}$

4. Which of the following numbers is a prime number? (Lesson 5.5)

   Ⓐ 1

   Ⓑ 11

   Ⓒ 21

   Ⓓ 51

5. A teacher has a bag of 100 unit cubes. She gives an equal number of cubes to each of the 7 groups in her class. She gives each group as many cubes as she can. How many unit cubes are left over?

   (Lesson 4.8)

   Ⓐ 1

   Ⓑ 2

   Ⓒ 3

   Ⓓ 6

6. Jessie sorted the coins in her bank. She made 7 stacks of 6 dimes and 8 stacks of 5 nickels. She then found 1 dime and 1 nickel. How many dimes and nickels does Jessie have in all? (Lesson 2.12)

   Ⓐ 84

   Ⓑ 82

   Ⓒ 80

   Ⓓ 28

Add Fractions Using Models

Find the sum. Use fraction strips to help.

1. $\frac{2}{6} + \frac{1}{6} =$ _____ $\dfrac{3}{6}$

$\frac{2}{6}$ $\frac{1}{6}$

2. $\frac{4}{10} + \frac{5}{10} =$ _____

3. $\frac{1}{3} + \frac{2}{3} =$ _____

4. $\frac{2}{4} + \frac{1}{4} =$ _____

5. $\frac{2}{12} + \frac{4}{12} =$ _____

6. $\frac{1}{6} + \frac{2}{6} =$ _____

7. $\frac{3}{12} + \frac{9}{12} =$ _____

8. $\frac{3}{8} + \frac{4}{8} =$ _____

9. $\frac{3}{4} + \frac{1}{4} =$ _____

10. $\frac{1}{5} + \frac{2}{5} =$ _____

Problem Solving REAL WORLD

11. Lola walks $\frac{4}{10}$ mile to her friend's house. Then she walks $\frac{5}{10}$ mile to the store. How far does she walk in all?

12. Evan eats $\frac{1}{8}$ of a pan of lasagna and his brother eats $\frac{2}{8}$ of it. What fraction of the pan of lasagna do they eat in all?

_____

_____

13. Jacqueline buys $\frac{2}{4}$ yard of green ribbon and $\frac{1}{4}$ yard of pink ribbon. How many yards of ribbon does she buy in all?

14. Shu mixes $\frac{2}{3}$ pound of peanuts with $\frac{1}{3}$ pound of almonds. How many pounds of nuts does Shu mix in all?

_____

_____

Lesson Check

1. Mary Jane has $\frac{3}{8}$ of a medium pizza left. Hector has $\frac{2}{8}$ of another medium pizza left. How much pizza do they have altogether?

Ⓐ $\frac{1}{8}$ Ⓒ $\frac{5}{8}$

Ⓑ $\frac{4}{8}$ Ⓓ $\frac{6}{8}$

2. Jeannie ate $\frac{1}{4}$ of an apple. Kelly ate $\frac{2}{4}$ of the apple. How much did they eat in all?

Ⓐ $\frac{1}{4}$ Ⓒ $\frac{3}{8}$

Ⓑ $\frac{2}{8}$ Ⓓ $\frac{3}{4}$

Spiral Review

3. Karen is making 14 different kinds of greeting cards. She is making 12 of each kind. How many greeting cards is she making? **(Lesson 2.10)**

Ⓐ 120

Ⓑ 132

Ⓒ 156

Ⓓ 168

4. Jefferson works part time and earns $1,520 in four weeks. How much does he earn each week? **(Lesson 4.11)**

Ⓐ $305

Ⓑ $350

Ⓒ $380

Ⓓ $385

5. By installing efficient water fixtures, the average American can reduce water use to about 45 gallons of water per day. Using such water fixtures, about how many gallons of water would the average American use in December? **(Lesson 3.2)**

Ⓐ about 1,200 gallons

Ⓑ about 1,500 gallons

Ⓒ about 1,600 gallons

Ⓓ about 2,000 gallons

6. Collin is making a bulletin board and note center. He is using square cork tiles and square dry-erase tiles. One of every 3 squares will be a cork square. If he uses 12 squares for the center, how many will be cork squares? **(Lesson 6.5)**

Ⓐ 3

Ⓑ 4

Ⓒ 6

Ⓓ 8

Subtract Fractions Using Models

Subtract. Use fraction strips to help.

1. $\dfrac{4}{5} - \dfrac{1}{5} = \dfrac{3}{5}$

1

$\frac{1}{5}$	$\frac{1}{5}$	$\frac{1}{5}$	$\frac{1}{5}$

2. $\dfrac{3}{4} - \dfrac{1}{4} = $ _____

1

$\frac{2}{4}$

3. $\dfrac{5}{6} - \dfrac{1}{6} = $ _____

4. $\dfrac{7}{8} - \dfrac{1}{8} = $ _____

5. $1 - \dfrac{2}{3} = $ _____

6. $\dfrac{8}{10} - \dfrac{2}{10} = $ _____

7. $\dfrac{3}{4} - \dfrac{1}{4} = $ _____

8. $\dfrac{7}{6} - \dfrac{5}{6} = $ _____

Problem Solving REAL WORLD

Use the table for 9 and 10.

9. Ena is making trail mix. She buys the items shown in the table. How many more pounds of pretzels than raisins does she buy?

_____

10. How many more pounds of granola than banana chips does she buy?

_____

Item	Weight (in pounds)
Pretzels	$\frac{7}{8}$
Peanuts	$\frac{4}{8}$
Raisins	$\frac{2}{8}$
Banana Chips	$\frac{3}{8}$
Granola	$\frac{5}{8}$

Lesson Check

1. Lee reads for $\frac{3}{4}$ hour in the morning and $\frac{2}{4}$ hour in the afternoon. How much longer does Lee read in the morning than in the afternoon?

   (A) 5 hours

   (B) $\frac{5}{4}$ hours

   (C) $\frac{4}{4}$ hour

   (D) $\frac{1}{4}$ hour

2. Which equation does the model below represent?

   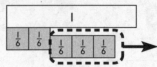

   (A) $\frac{3}{6} - \frac{2}{6} = \frac{1}{6}$

   (B) $\frac{2}{6} - \frac{1}{6} = \frac{1}{6}$

   (C) $\frac{5}{6} - \frac{3}{6} = \frac{2}{6}$

   (D) $1 - \frac{3}{6} = \frac{3}{6}$

Spiral Review

3. A city received 2 inches of rain each day for 3 days. The meteorologist said that if the rain had been snow, each inch of rain would have been 10 inches of snow. How much snow would that city have received in the 3 days? (Lesson 2.8)

   (A) 20 inches

   (B) 30 inches

   (C) 50 inches

   (D) 60 inches

4. At a party there were four large submarine sandwiches, all the same size. During the party, $\frac{2}{3}$ of the chicken sandwich, $\frac{3}{4}$ of the tuna sandwich, $\frac{7}{12}$ of the roast beef sandwich, and $\frac{5}{6}$ of the veggie sandwich were eaten. Which sandwich had the least amount left? (Lesson 6.8)

   (A) chicken

   (B) tuna

   (C) roast beef

   (D) veggie

5. Deena uses $\frac{3}{8}$ cup milk and $\frac{2}{8}$ cup oil in a recipe. How much liquid does she use in all? (Lesson 7.3)

   (A) $\frac{1}{8}$ cup

   (B) $\frac{5}{8}$ cup

   (C) $\frac{6}{8}$ cup

   (D) 5 cups

6. In the car lot, $\frac{4}{12}$ of the cars are white and $\frac{3}{12}$ of the cars are blue. What fraction of the cars in the lot are either white or blue? (Lesson 7.3)

   (A) $\frac{1}{12}$

   (B) $\frac{7}{24}$

   (C) $\frac{7}{12}$

   (D) 7

Name _____

Add and Subtract Fractions

Find the sum or difference.

1. $\dfrac{4}{12} + \dfrac{8}{12} = \dfrac{12}{12}$ _____

$$\frac{4}{12} \qquad \frac{8}{12}$$

2. $\dfrac{3}{6} - \dfrac{1}{6} =$ _____

$$\frac{2}{6}$$

3. $\dfrac{4}{5} - \dfrac{3}{5} =$ _____

4. $\dfrac{6}{10} + \dfrac{3}{10} =$ _____

5. $1 - \dfrac{3}{8} =$ _____

6. $\dfrac{1}{4} + \dfrac{2}{4} =$ _____

7. $\dfrac{9}{12} - \dfrac{5}{12} =$ _____

8. $\dfrac{5}{6} - \dfrac{2}{6} =$ _____

9. $\dfrac{2}{3} + \dfrac{1}{3} =$ _____

Problem Solving REAL WORLD

Use the table for 10 and 11.

10. Guy finds how far his house is from several locations and makes the table shown. How much farther away from Guy's house is the library than the cafe?

_____

11. If Guy walks from his house to school and back, how far does he walk?

_____

Distance from Guy's House	
Location	**Distance (in miles)**
Library	$\dfrac{9}{10}$
School	$\dfrac{5}{10}$
Store	$\dfrac{7}{10}$
Cafe	$\dfrac{4}{10}$
Yogurt Shop	$\dfrac{6}{10}$

Lesson Check

1. Mr. Angulo buys $\frac{5}{8}$ pound of red grapes and $\frac{3}{8}$ pound of green grapes. How many pounds of grapes did Mr. Angulo buy in all?

   (A) $\frac{1}{8}$ pound

   (B) $\frac{2}{8}$ pound

   (C) 1 pound

   (D) 2 pounds

2. Which equation does the model below represent?

   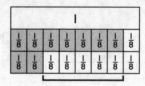

   (A) $\frac{7}{8} + \frac{2}{8} = \frac{9}{8}$

   (B) $\frac{5}{8} - \frac{2}{8} = \frac{3}{8}$

   (C) $\frac{8}{8} - \frac{5}{8} = \frac{3}{8}$

   (D) $\frac{7}{8} - \frac{2}{8} = \frac{5}{8}$

Spiral Review

3. There are 6 muffins in a package. How many packages will be needed to feed 48 people if each person has 2 muffins? (Lesson 4.12)

   (A) 4   (C) 16

   (B) 8   (D) 24

4. Camp Oaks gets 32 boxes of orange juice and 56 boxes of apple juice. Each shelf in the cupboard can hold 8 boxes of juice. What is the least number of shelves needed for all the juice boxes? (Lesson 4.12)

   (A) 4   (C) 11

   (B) 7   (D) 88

5. A machine makes 18 parts each hour. If the machine operates 24 hours a day, how many parts can it make in one day? (Lesson 3.6)

   (A) 302

   (B) 332

   (C) 362

   (D) 432

6. Which equation does the model below represent? (Lesson 7.4)

   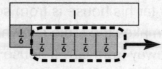

   (A) $\frac{5}{6} - \frac{4}{6} = \frac{1}{6}$

   (B) $\frac{4}{5} - \frac{1}{5} = \frac{3}{5}$

   (C) $\frac{5}{5} - \frac{4}{5} = \frac{1}{5}$

   (D) $\frac{6}{6} - \frac{4}{6} = \frac{2}{6}$

Name _____

Rename Fractions and Mixed Numbers

Write the mixed number as a fraction.

1. $2\frac{3}{5}$

Think: Find $\frac{5}{5} + \frac{5}{5} + \frac{3}{5}$.

$\frac{13}{5}$

_____

2. $4\frac{1}{3}$

_____

3. $1\frac{2}{5}$

_____

4. $3\frac{2}{3}$

_____

5. $4\frac{1}{8}$

_____

6. $1\frac{7}{10}$

_____

7. $5\frac{1}{2}$

_____

8. $2\frac{3}{8}$

_____

Write the fraction as a mixed number.

9. $\frac{31}{6}$

_____

10. $\frac{20}{10}$

_____

11. $\frac{15}{8}$

_____

12. $\frac{13}{6}$

_____

13. $\frac{23}{10}$

_____

14. $\frac{19}{5}$

_____

15. $\frac{11}{3}$

_____

16. $\frac{9}{2}$

_____

Problem Solving

17. A recipe calls for $2\frac{2}{4}$ cups of raisins, but Julie only has a $\frac{1}{4}$-cup measuring cup. How many $\frac{1}{4}$ cups does Julie need to measure out $2\frac{2}{4}$ cups of raisins?

18. If Julie needs $3\frac{1}{4}$ cups of oatmeal, how many $\frac{1}{4}$ cups of oatmeal will she use?

_____   _____

Lesson Check

1. Which of the following is equivalent to $\frac{16}{3}$?

   (A) $3\frac{1}{5}$ (C) $5\frac{1}{3}$

   (B) $3\frac{2}{5}$ (D) $5\frac{6}{3}$

2. Stacey filled her $\frac{1}{2}$-cup measuring cup seven times to have enough flour for a cake recipe. How much flour does the cake recipe call for?

   (A) 3 cups   (C) 4 cups

   (B) $3\frac{1}{2}$ cups   (D) $4\frac{1}{2}$ cups

Spiral Review

3. Becki put some stamps into her stamp collection book. She put 14 stamps on each page. If she completely filled 16 pages, how many stamps did she put in the book? **(Lesson 3.5)**

   (A) 224

   (B) 240

   (C) 272

   (D) 275

4. Brian is driving 324 miles to visit some friends. He wants to get there in 6 hours. How many miles does he need to drive each hour? **(Lesson 4.10)**

   (A) 48 miles

   (B) 50 miles

   (C) 52 miles

   (D) 54 miles

5. During a bike challenge, riders have to collect various colored ribbons. Each $\frac{1}{2}$ mile they collect a red ribbon, each $\frac{1}{8}$ mile they collect a green ribbon, and each $\frac{1}{4}$ mile they collect a blue ribbon. Which colors of ribbons will be collected at the $\frac{3}{4}$ mile marker? **(Lesson 6.5)**

   (A) red and green

   (B) red and blue

   (C) green and blue

   (D) red, green, and blue

6. Stephanie had $\frac{7}{8}$ pound of bird seed. She used $\frac{3}{8}$ pound to fill a bird feeder. How much bird seed does Stephanie have left? **(Lesson 7.5)**

   (A) $\frac{3}{8}$ pound

   (B) $\frac{4}{8}$ pound

   (C) 1 pound

   (D) $\frac{10}{8}$ pounds

Name _____

Add and Subtract Mixed Numbers

Find the sum. Write the sum as a mixed number,
so the fractional part is less than 1.

1. $6\frac{4}{5}$
    $+3\frac{3}{5}$
    $\overline{9\frac{7}{5}} = 10\frac{2}{5}$

2. $4\frac{1}{2}$
    $+2\frac{1}{2}$

3. $2\frac{2}{3}$
    $+3\frac{2}{3}$

4. $6\frac{4}{5}$
    $+7\frac{4}{5}$

5. $9\frac{3}{6}$
    $+2\frac{2}{6}$

6. $8\frac{4}{12}$
    $+3\frac{6}{12}$

7. $4\frac{3}{8}$
    $+1\frac{5}{8}$

8. $9\frac{5}{10}$
    $+6\frac{3}{10}$

Find the difference.

9. $6\frac{7}{8}$
    $-4\frac{3}{8}$

10. $4\frac{2}{3}$
    $-3\frac{1}{3}$

11. $6\frac{4}{5}$
    $-3\frac{3}{5}$

12. $7\frac{3}{4}$
    $-2\frac{1}{4}$

Problem Solving 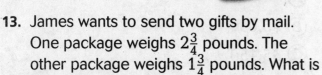 REAL WORLD

13. James wants to send two gifts by mail.
    One package weighs $2\frac{3}{4}$ pounds. The
    other package weighs $1\frac{3}{4}$ pounds. What is
    the total weight of the packages?

14. Tierra bought $4\frac{3}{8}$ yards blue ribbon and
    $2\frac{1}{8}$ yards yellow ribbon for a craft project.
    How much more blue ribbon than yellow
    ribbon did Tierra buy?

_____

_____

Lesson Check

1. Brad has two lengths of copper pipe to fit together. One has a length of $2\frac{5}{12}$ feet and the other has a length of $3\frac{7}{12}$ feet. How many feet of pipe does he have in all?

  Ⓐ 5 feet   Ⓒ $5\frac{10}{12}$ feet

  Ⓑ $5\frac{6}{12}$ feet   Ⓓ 6 feet

2. A pattern calls for $2\frac{1}{4}$ yards of material and $1\frac{1}{4}$ yards of lining. How much total fabric is needed?

  Ⓐ $1\frac{2}{4}$ yards   Ⓒ $3\frac{1}{4}$ yards

  Ⓑ 3 yards   Ⓓ $3\frac{2}{4}$ yards

Spiral Review

3. Shanice has 23 baseball trading cards of star players. She agrees to sell them for $16 each. How much will she get for the cards? **(Lesson 3.3)**

  Ⓐ $258

  Ⓑ $358

  Ⓒ $368

  Ⓓ $468

4. Nanci is volunteering at the animal shelter. She wants to spend an equal amount of time playing with each dog. She has 145 minutes to play with all 7 dogs. About how much time can she spend with each dog? **(Lesson 4.1)**

  Ⓐ about 10 minutes

  Ⓑ about 20 minutes

  Ⓒ about 25 minutes

  Ⓓ about 26 minutes

5. Frieda has 12 red apples and 15 green apples. She is going to share the apples equally among 8 people and keep any extra apples for herself. How many apples will Frieda keep for herself? **(Lesson 4.3)**

  Ⓐ 3

  Ⓑ 4

  Ⓒ 6

  Ⓓ 7

6. The Lynch family bought a house for $75,300. A few years later, they sold the house for $80,250. How much greater was the selling price than the purchase price? **(Lesson 1.8)**

  Ⓐ $4,950

  Ⓑ $5,050

  Ⓒ $5,150

  Ⓓ $5,950

Name _____

Record Subtraction with Renaming

Find the difference.

1. $5\frac{1}{3} \rightarrow 4\frac{4}{3}$
    $-3\frac{2}{3} \rightarrow 3\frac{2}{3}$
    _____
    $1\frac{2}{3}$

2. 6
    $-3\frac{2}{5}$
    _____

3. $5\frac{1}{4}$
    $-2\frac{3}{4}$
    _____

4. $9\frac{3}{8}$
    $-8\frac{7}{8}$
    _____

5. $12\frac{3}{10}$
    $-7\frac{7}{10}$
    _____

6. $8\frac{1}{6}$
    $-3\frac{5}{6}$
    _____

7. $7\frac{3}{5}$
    $-4\frac{4}{5}$
    _____

8. $10\frac{1}{2}$
    $-8\frac{1}{2}$
    _____

9. $7\frac{1}{6}$
    $-2\frac{5}{6}$
    _____

10. $9\frac{3}{12}$
    $-4\frac{7}{12}$
    _____

11. $9\frac{1}{10}$
    $-8\frac{7}{10}$
    _____

12. $9\frac{1}{3}$
    $-2\frac{2}{3}$
    _____

13. $3\frac{1}{4}$
    $-1\frac{3}{4}$
    _____

14. $4\frac{5}{8}$
    $-1\frac{7}{8}$
    _____

15. $5\frac{1}{12}$
    $-3\frac{8}{12}$
    _____

16. 7
    $-1\frac{3}{5}$
    _____

Problem Solving REAL WORLD

17. Alicia buys a 5-pound bag of rocks for a fish tank. She uses $1\frac{1}{8}$ pounds for a small fish bowl. How much is left?

18. Xavier made 25 pounds of roasted almonds for a fair. He has $3\frac{1}{2}$ pounds left at the end of the fair. How many pounds of roasted almonds did he sell at the fair?

Lesson Check

1. Reggie is making a double-layer cake. The recipe for the first layer calls for $2\frac{1}{4}$ cups sugar. The recipe for the second layer calls for $1\frac{1}{4}$ cups sugar. Reggie has 5 cups of sugar. How much will he have left after making both recipes?

   (A) $1\frac{1}{4}$ cups   (C) $2\frac{1}{4}$ cups

   (B) $1\frac{2}{4}$ cups   (D) $2\frac{2}{4}$ cups

2. Kate has $4\frac{3}{8}$ yards of fabric and needs $2\frac{7}{8}$ yards to make a skirt. How much extra fabric will she have left after making the skirt?

   (A) $2\frac{4}{8}$ yards   (C) $1\frac{4}{8}$ yards

   (B) $2\frac{2}{8}$ yards   (D) $1\frac{2}{8}$ yards

Spiral Review

3. Paulo has 128 glass beads to use to decorate picture frames. He wants to use the same number of beads on each frame. If he decorates 8 picture frames, how many beads will he put on each frame? (Lesson 4.8)

   (A) 6

   (B) 7

   (C) 14

   (D) 16

4. Madison is making party favors. She wants to make enough favors so each guest gets the same number of favors. She knows there will be 6 or 8 guests at the party. What is the least number of party favors Madison should make? (Lesson 5.4)

   (A) 18

   (B) 24

   (C) 30

   (D) 32

5. A shuttle bus makes 4 round-trips between two shopping centers each day. The bus holds 24 people. If the bus is full on each one-way trip, how many passengers are carried by the bus each day? (Lesson 2.10)

   (A) 96

   (B) 162

   (C) 182

   (D) 192

6. To make a fruit salad, Marvin mixes $1\frac{3}{4}$ cups of diced peaches with $2\frac{1}{4}$ cups of diced pears. How many cups of peaches and pears are in the fruit salad? (Lesson 7.7)

   (A) 4 cups

   (B) $3\frac{2}{4}$ cups

   (C) $3\frac{1}{4}$ cups

   (D) 3 cups

Name _____

Fractions and Properties of Addition

Use the properties and mental math to find the sum.

1. $5\frac{1}{3} + \left(2\frac{2}{3} + 1\frac{1}{3}\right)$

$5\frac{1}{3} + (4)$

$9\frac{1}{3}$
_____

2. $10\frac{1}{8} + \left(3\frac{5}{8} + 2\frac{7}{8}\right)$
_____

3. $8\frac{1}{5} + \left(3\frac{2}{5} + 5\frac{4}{5}\right)$
_____

4. $6\frac{3}{4} + \left(4\frac{2}{4} + 5\frac{1}{4}\right)$
_____

5. $\left(6\frac{3}{6} + 10\frac{4}{6}\right) + 9\frac{2}{6}$
_____

6. $\left(6\frac{2}{5} + 1\frac{4}{5}\right) + 3\frac{1}{5}$
_____

7. $7\frac{7}{8} + \left(3\frac{1}{8} + 1\frac{1}{8}\right)$
_____

8. $14\frac{1}{10} + \left(20\frac{2}{10} + 15\frac{7}{10}\right)$
_____

9. $\left(13\frac{2}{12} + 8\frac{7}{12}\right) + 9\frac{5}{12}$
_____

Problem Solving  REAL WORLD

10. Nate's classroom has three tables of different lengths. One has a length of $4\frac{1}{2}$ feet, another has a length of 4 feet, and a third has a length of $2\frac{1}{2}$ feet. What is the length of all three tables when pushed end to end?

_____

11. Mr. Warren uses $2\frac{1}{4}$ bags of mulch for his garden and another $4\frac{1}{4}$ bags for his front yard. He also uses $\frac{3}{4}$ bag around a fountain. How many total bags of mulch does Mr. Warren use?

_____

Lesson Check

1. A carpenter cut a board into three pieces. One piece was $2\frac{5}{6}$ feet long. The second piece was $3\frac{1}{6}$ feet long. The third piece was $1\frac{5}{6}$ feet long. How long was the board?

   (A) $6\frac{5}{6}$ feet

   (B) $7\frac{1}{6}$ feet

   (C) $7\frac{5}{6}$ feet

   (D) $8\frac{1}{6}$ feet

2. Harry works at an apple orchard. He picked $45\frac{7}{8}$ pounds of apples on Monday. He picked $42\frac{3}{8}$ pounds of apples on Wednesday. He picked $54\frac{1}{8}$ pounds of apples on Friday. How many pounds of apples did Harry pick those three days?

   (A) $132\frac{3}{8}$ pounds

   (B) $141\frac{3}{8}$ pounds

   (C) $142\frac{1}{8}$ pounds

   (D) $142\frac{3}{8}$ pounds

Spiral Review

3. There were 6 oranges in the refrigerator. Joey and his friends ate $3\frac{2}{3}$ oranges. How many oranges were left? **(Lesson 7.8)**

   (A) $2\frac{1}{3}$ oranges

   (B) $2\frac{2}{3}$ oranges

   (C) $3\frac{1}{3}$ oranges

   (D) $9\frac{2}{3}$ oranges

4. Darlene was asked to identify which of the following numbers is prime. Which number should she choose? **(Lesson 5.5)**

   (A) 2

   (B) 12

   (C) 21

   (D) 39

5. A teacher has 100 chairs to arrange for an assembly. Which of the following is NOT a way the teacher could arrange the chairs? **(Lesson 5.2)**

   (A) 10 rows of 10 chairs

   (B) 8 rows of 15 chairs

   (C) 5 rows of 20 chairs

   (D) 4 rows of 25 chairs

6. Nic bought 28 folding chairs for $16 each. How much money did Nic spend on chairs? **(Lesson 3.5)**

   (A) $196

   (B) $348

   (C) $448

   (D) $600

Name _____

Problem Solving • Multistep Fraction Problems

Read each problem and solve.

1. Each child in the Smith family was given an orange cut into 8 equal sections. Each child ate $\frac{5}{8}$ of the orange. After combining the leftover sections, Mrs. Smith noted that there were exactly 3 full oranges left. How many children are in the Smith family?

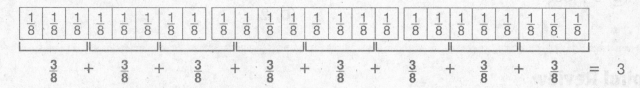

$$\frac{3}{8} + \frac{3}{8} + \frac{3}{8} + \frac{3}{8} + \frac{3}{8} + \frac{3}{8} + \frac{3}{8} + \frac{3}{8} = 3$$

There are 8 addends, so there are 8 children in the Smith family.

_____ 8 children _____

2. Val walks $2\frac{3}{5}$ miles each day. Bill runs 10 miles once every 4 days. In 4 days, who covers the greater distance?

_____

3. Chad buys peanuts in 2-pound bags. He repackages them into bags that hold $\frac{5}{6}$ pound of peanuts. How many 2-pound bags of peanuts should Chad buy so that he can fill the $\frac{5}{6}$-pound bags without having any peanuts left over?

_____

4. A carpenter has several boards of equal length. He cuts $\frac{3}{5}$ of each board. After cutting the boards, the carpenter notices that he has enough pieces left over to make up the same length as 4 of the original boards. How many boards did the carpenter start with?

_____

Lesson Check

1. Karyn cuts a length of ribbon into 4 equal pieces, each $1\frac{1}{4}$ feet long. How long was the ribbon?

   (A) 4 feet

   (B) $4\frac{1}{4}$ feet

   (C) 5 feet

   (D) $5\frac{1}{4}$ feet

2. Several friends each had $\frac{2}{5}$ of a bag of peanuts left over from the baseball game. They realized that they could have bought 2 fewer bags of peanuts between them. How many friends went to the game?

   (A) 6

   (B) 5

   (C) 4

   (D) 2

Spiral Review

3. A frog made three jumps. The first was $12\frac{5}{6}$ inches. The second jump was $8\frac{3}{6}$ inches. The third jump was $15\frac{1}{6}$ inches. What was the total distance the frog jumped? **(Lesson 7.9)**

   (A) $35\frac{3}{6}$ inches

   (B) $36\frac{1}{6}$ inches

   (C) $36\frac{3}{6}$ inches

   (D) $38\frac{1}{6}$ inches

4. LaDanian wants to write the fraction $\frac{4}{6}$ as a sum of unit fractions. Which expression should he write? **(Lesson 7.2)**

   (A) $\frac{1}{6} + \frac{1}{6} + \frac{1}{6} + \frac{1}{6}$

   (B) $\frac{2}{6} + \frac{2}{6}$

   (C) $\frac{3}{6} + \frac{1}{6}$

   (D) $\frac{1}{6} + \frac{1}{6} + \frac{2}{6}$

5. Greta made a design with squares. She colored 8 out of the 12 squares blue. What fraction of the squares did she color blue? **(Lesson 6.3)**

   (A) $\frac{1}{4}$

   (B) $\frac{1}{3}$

   (C) $\frac{2}{3}$

   (D) $\frac{3}{4}$

6. The teacher gave this pattern to the class: the first term is 5 and the rule is *add* 4, *subtract* 1. Each student says one number. The first student says 5. Victor is tenth in line. What number should Victor say? **(Lesson 5.6)**

   (A) 17

   (B) 19

   (C) 20

   (D) 21

Name _____

Chapter 7 Extra Practice

Lesson 7.1

Use the model to write an equation.

1. 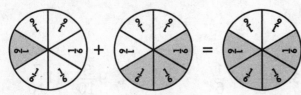 2.

_____   _____

Use the model to solve the equation.

3. $\dfrac{3}{10} + \dfrac{5}{10} =$ _____

4. $\dfrac{7}{12} - \dfrac{6}{12} =$ _____

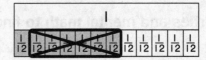

Lesson 7.2

Write the fraction as a sum of unit fractions.

1. $\dfrac{2}{3} =$ _____

2. $\dfrac{3}{10} =$ _____

3. $\dfrac{4}{6} =$ _____

4. $\dfrac{5}{12} =$ _____

Lessons 7.3 - 7.5

Find the sum or difference. Use fraction strips to help.

1. $\dfrac{3}{8} + \dfrac{2}{8} =$ _____

2. $\dfrac{4}{5} + \dfrac{1}{5} =$ _____

3. $\dfrac{6}{10} + \dfrac{1}{10} =$ _____

4. $\dfrac{5}{6} - \dfrac{4}{6} =$ _____

5. $\dfrac{3}{4} - \dfrac{1}{4} =$ _____

6. $1 - \dfrac{7}{12} =$ _____

7. $\dfrac{7}{10} - \dfrac{3}{10} =$ _____

8. $\dfrac{2}{6} + \dfrac{4}{6} =$ _____

9. $\dfrac{5}{8} - \dfrac{4}{8} =$ _____

Lesson 7.6

Write each mixed number as a fraction and each fraction as a mixed number.

1. $4\frac{2}{3} =$ _____

2. $6\frac{1}{4} =$ _____

3. $\frac{11}{3} =$ _____

4. $\frac{16}{15} =$ _____

Lessons 7.7 - 7.8

Find the sum or difference.

1. $3\frac{1}{4} + 2\frac{3}{4}$

2. $1\frac{5}{12} + 2\frac{1}{12}$

3. $9\frac{5}{6} - 7\frac{1}{6}$

4. $9\frac{3}{10} - 1\frac{7}{10}$

_____   _____   _____   _____

Lesson 7.9

Use the properties and mental math to find the sum.

1. $\left(1\frac{1}{4} + 4\right) + 2\frac{3}{4}$

2. $\frac{3}{5} + \left(90\frac{2}{5} + 10\right)$

3. $3\frac{2}{6} + \left(2\frac{1}{6} + \frac{4}{6}\right)$

4. $\left(\frac{5}{8} + 2\frac{3}{8}\right) + 1\frac{3}{8}$

_____   _____   _____   _____

Lesson 7.10

1. Adrian jogs $\frac{3}{4}$ mile each morning. How many days will it take him to jog 3 miles?

_____

2. Trail mix is sold in 1-pound bags. Mary will buy some trail mix and re-package it so that each of the 15 members of her hiking club gets one $\frac{2}{5}$-pound bag. How many 1-pound bags of trail mix should Mary buy to have enough trail mix without leftovers?

_____

School-Home Letter

© Houghton Mifflin Harcourt Publishing Company

Vocabulary

mixed number A number represented by a whole number and a fraction

multiple A number that is the product of a given number and a counting number

unit fraction A fraction that has 1 as its top number or numerator

Dear Family,

During the next few weeks, our math class will be learning how to multiply fractions and mixed numbers by whole numbers. We will learn to write a fraction as a product of a whole number and a unit fraction, and to find multiples of unit fractions.

You can expect to see homework that provides practice multiplying fractions and whole numbers with and without using models.

Here is a sample of how your child will be taught to use a number line to find multiples of a fraction.

🔑 MODEL Use a Number Line to Write Multiples of Fractions

Write $3 \times \frac{3}{4}$ as the product of a whole number and a unit fraction.

STEP 1

Start at 0. Draw jumps to find multiples of $\frac{3}{4}$: $\frac{3}{4}, \frac{6}{4}, \frac{9}{4}$.

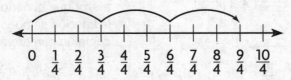

$$0 \quad \frac{1}{4} \quad \frac{2}{4} \quad \frac{3}{4} \quad \frac{4}{4} \quad \frac{5}{4} \quad \frac{6}{4} \quad \frac{7}{4} \quad \frac{8}{4} \quad \frac{9}{4} \quad \frac{10}{4}$$

STEP 2

Write the multiple as a product of a whole number and a unit fraction.

So, $3 \times \frac{3}{4} = \frac{9}{4} = 9 \times \frac{1}{4}$.

Tips

Renaming as a Mixed Number

When the numerator is greater than the denominator, the fraction can be renamed as a mixed number.

$$\frac{9}{4} = \frac{4}{4} + \frac{4}{4} + \frac{1}{4}$$
$$= 2 + \frac{1}{4}$$
$$= 2\frac{1}{4}$$

Activity

Use everyday situations, such as cooking and measures to help your child practice fraction multiplication.

Carta
para la casa

© Houghton Mifflin Harcourt Publishing Company

Vocabulary

fracción unitaria Una fracción que tiene al 1 como numerador, es decir, arriba de la barra

múltiplo Un número que es el producto de cierto número y un número positivo distinto de cero

número mixto Un número que se representa por un número entero y una fracción

Querida familia,

Durante las próximas semanas, en la clase de matemáticas aprenderemos a multiplicar fracciones y números mixtos por números enteros. También aprenderemos a escribir fracciones como el producto de un número entero y una fracción unitaria y a hallar múltiplos de fracciones unitarias.

Llevaré a casa tareas para practicar la multiplicación de fracciones y números enteros usando modelos y sin modelos.

Este es un ejemplo de cómo vamos a usar una recta numérica para hallar los múltiplos de una fracción.

🔑 MODELO Usar una recta numérica para escribir múltiplos de fracciones

Escribe $3 \times \frac{3}{4}$ como el producto de un número entero y una fracción unitaria.

PASO 1

Comienza en 0. Dibuja saltos para hallar los múltiplos de $\frac{3}{4}$: $\frac{3}{4}, \frac{6}{4}, \frac{9}{4}$

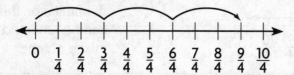

$$0 \quad \frac{1}{4} \quad \frac{2}{4} \quad \frac{3}{4} \quad \frac{4}{4} \quad \frac{5}{4} \quad \frac{6}{4} \quad \frac{7}{4} \quad \frac{8}{4} \quad \frac{9}{4} \quad \frac{10}{4}$$

PASO 2

Escribe el múltiplo como el producto de un número entero y una fracción unitaria.

Por lo tanto, $3 \times \frac{3}{4} = \frac{9}{4} = 9 \times \frac{1}{4}$.

Pistas

Expresarlo como un número mixto

Cuando el numerador es mayor que el denominador, la fracción se puede expresar como un número mixto.

$$\frac{9}{4} = \frac{4}{4} + \frac{4}{4} + \frac{1}{4}$$
$$= 2 + \frac{1}{4}$$
$$= 2\frac{1}{4}$$

Actividad

Use situaciones de la vida diaria, como cocinar y medir para ayudar a su hijo o hija a practicar la multiplicación con fracciones.

Name _____

Multiples of Unit Fractions

Write the fraction as a product of a whole number and a unit fraction.

1. $\dfrac{5}{6} =$ $5 \times \dfrac{1}{6}$

2. $\dfrac{7}{8} =$ _____

3. $\dfrac{5}{3} =$ _____

4. $\dfrac{9}{10} =$ _____

5. $\dfrac{3}{4} =$ _____

6. $\dfrac{11}{12} =$ _____

7. $\dfrac{4}{6} =$ _____

8. $\dfrac{8}{20} =$ _____

9. $\dfrac{13}{100} =$ _____

List the next four multiples of the unit fraction.

10. $\dfrac{1}{5}$, ____, ____, ____, ____

11. $\dfrac{1}{8}$, ____, ____, ____, ____

Problem Solving  REAL WORLD

12. So far, Monica has read $\dfrac{5}{6}$ of a book. She has read the same number of pages each day for 5 days. What fraction of the book does Monica read each day?

_____

13. Nicholas buys $\dfrac{3}{8}$ pound of cheese. He puts the same amount of cheese on 3 sandwiches. How much cheese does Nicholas put on each sandwich?

_____

Lesson Check

1. Selena walks from home to school each morning and back home each afternoon. Altogether, she walks $\frac{2}{3}$ mile each day. How far does Selena live from school?

    Ⓐ $\frac{1}{3}$ mile

    Ⓑ $\frac{2}{3}$ mile

    Ⓒ $1\frac{1}{3}$ miles

    Ⓓ 2 miles

2. Will uses $\frac{3}{4}$ cup of olive oil to make 3 batches of salad dressing. How much oil does Will use for one batch of salad dressing?

    Ⓐ $\frac{1}{4}$ cup

    Ⓑ $\frac{1}{3}$ cup

    Ⓒ $2\frac{1}{4}$ cups

    Ⓓ 3 cups

Spiral Review

3. Liza bought $\frac{5}{8}$ pound of trail mix. She gives $\frac{1}{8}$ pound of trail mix to Michael. How much trail mix does Liza have left? **(Lesson 7.5)**

    Ⓐ $\frac{1}{8}$ pound

    Ⓑ $\frac{2}{8}$ pound

    Ⓒ $\frac{3}{8}$ pound

    Ⓓ $\frac{4}{8}$ pound

4. Leigh has a piece of rope that is $6\frac{2}{3}$ feet long. How do you write $6\frac{2}{3}$ as a fraction greater than 1? **(Lesson 7.6)**

    Ⓐ $\frac{11}{3}$

    Ⓑ $\frac{15}{3}$

    Ⓒ $\frac{20}{3}$

    Ⓓ $\frac{62}{3}$

5. Randy's house number is a composite number. Which of the following could be Randy's house number? **(Lesson 5.5)**

    Ⓐ 29

    Ⓑ 39

    Ⓒ 59

    Ⓓ 79

6. Mindy buys 12 cupcakes. Nine of the cupcakes have chocolate frosting and the rest have vanilla frosting. What fraction of the cupcakes have vanilla frosting? **(Lesson 6.3)**

    Ⓐ $\frac{1}{4}$

    Ⓑ $\frac{1}{3}$

    Ⓒ $\frac{2}{3}$

    Ⓓ $\frac{3}{4}$

Name _____

Multiples of Fractions

List the next four multiples of the fraction.

1. $\frac{3}{5}$, ____, ____, ____, ____

2. $\frac{2}{6}$, ____, ____, ____, ____

3. $\frac{4}{8}$, ____, ____, ____, ____

4. $\frac{5}{10}$, ____, ____, ____, ____

Write the product as the product of a whole number and a unit fraction.

5.

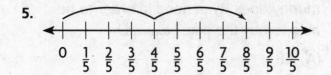

$2 \times \frac{4}{5} =$ _____

6.

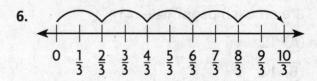

$5 \times \frac{2}{3} =$ _____

Problem Solving REAL WORLD

7. Jessica is making 2 loaves of banana bread. She needs $\frac{3}{4}$ cup of sugar for each loaf. Her measuring cup can only hold $\frac{1}{4}$ cup of sugar. How many times will Jessica need to fill the measuring cup in order to get enough sugar for both loaves of bread?

8. A group of four students is performing an experiment with salt. Each student must add $\frac{3}{8}$ teaspoon of salt to a solution. The group only has a $\frac{1}{8}$-teaspoon measuring spoon. How many times will the group need to fill the measuring spoon in order to perform the experiment?

_____   _____

Lesson Check

1. Eloise made a list of some multiples of $\frac{5}{8}$. Which of the following lists could be Eloise's list?

   (A) $\frac{5}{8}, \frac{10}{16}, \frac{15}{24}, \frac{20}{32}, \frac{25}{40}$

   (B) $\frac{5}{8}, \frac{10}{8}, \frac{15}{8}, \frac{20}{8}, \frac{25}{8}$

   (C) $\frac{5}{8}, \frac{6}{8}, \frac{7}{8}, \frac{8}{8}, \frac{9}{8}$

   (D) $\frac{1}{8}, \frac{2}{8}, \frac{3}{8}, \frac{4}{8}, \frac{5}{8}$

2. David is filling five $\frac{3}{4}$-quart bottles with a sports drink. His measuring cup only holds $\frac{1}{4}$ quart. How many times will David need to fill the measuring cup in order to fill the 5 bottles?

   (A) 5

   (B) 10

   (C) 15

   (D) 20

Spiral Review

3. Ira has 128 stamps in his stamp album. He has the same number of stamps on each of the 8 pages. How many stamps are on each page? **(Lesson 4.11)**

   (A) 12

   (B) 14

   (C) 16

   (D) 18

4. Ryan is saving up for a bike that costs $198. So far, he has saved $15 per week for the last 12 weeks. How much more money does Ryan need in order to be able to buy the bike? **(Lesson 3.7)**

   (A) $8

   (B) $18

   (C) $48

   (D) $180

5. Tina buys $3\frac{7}{8}$ yards of material at the fabric store. She uses it to make a skirt. Afterward, she has $1\frac{3}{8}$ yards of the fabric leftover. How many yards of material did Tina use? **(Lesson 7.7)**

   (A) $1\frac{4}{8}$ yards

   (B) $2\frac{1}{8}$ yards

   (C) $2\frac{4}{8}$ yards

   (D) $5\frac{2}{8}$ yards

6. Which list shows the fractions in order from **least** to **greatest**? **(Lesson 6.8)**

   (A) $\frac{2}{3}, \frac{3}{4}, \frac{7}{12}$

   (B) $\frac{7}{12}, \frac{3}{4}, \frac{2}{3}$

   (C) $\frac{3}{4}, \frac{2}{3}, \frac{7}{12}$

   (D) $\frac{7}{12}, \frac{2}{3}, \frac{3}{4}$

Multiply a Fraction by a Whole Number Using Models

Multiply.

1. $2 \times \dfrac{5}{6} =$ _____ $\dfrac{10}{6}$ _____

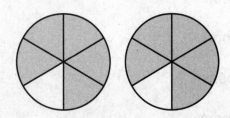

2. $3 \times \dfrac{2}{5} =$ _____

3. $7 \times \dfrac{3}{10} =$ _____

4. $3 \times \dfrac{5}{12} =$ _____

5. $6 \times \dfrac{3}{4} =$ _____

6. $4 \times \dfrac{2}{8} =$ _____

7. $5 \times \dfrac{2}{3} =$ _____

8. $2 \times \dfrac{7}{8} =$ _____

9. $6 \times \dfrac{4}{5} =$ _____

Problem Solving

10. Matthew walks $\dfrac{5}{8}$ mile to the bus stop each morning. How far will he walk in 5 days?

11. Emily uses $\dfrac{2}{3}$ cup of milk to make one batch of muffins. How many cups of milk will Emily use if she makes 3 batches of muffins?

_____

_____

Lesson Check

1. Aleta's puppy gained $\frac{3}{8}$ pound each week for 4 weeks. Altogether, how much weight did the puppy gain during the 4 weeks?

   (A) $\frac{8}{12}$ pound

   (B) $1\frac{2}{8}$ pounds

   (C) $\frac{12}{8}$ pounds

   (D) $4\frac{3}{8}$ pounds

2. Pedro mixes $\frac{3}{4}$ teaspoon of plant food into each gallon of water. How many teaspoons of plant food should Pedro mix into 5 gallons of water?

   (A) $\frac{3}{20}$ teaspoon

   (B) $\frac{4}{15}$ teaspoon

   (C) $\frac{8}{4}$ teaspoons

   (D) $\frac{15}{4}$ teaspoons

Spiral Review

3. Ivana has $\frac{3}{4}$ pound of hamburger meat. She makes 3 hamburger patties. Each patty weighs the same amount. How much does each hamburger patty weigh? (Lesson 8.1)

   (A) $\frac{1}{4}$ pound

   (B) $\frac{1}{3}$ pound

   (C) $2\frac{1}{4}$ pounds

   (D) 3 pounds

4. Which of the following expressions is NOT equal to $\frac{7}{10}$? (Lesson 7.2)

   (A) $\frac{5}{10} + \frac{1}{10} + \frac{1}{10}$

   (B) $\frac{2}{10} + \frac{2}{10} + \frac{3}{10}$

   (C) $\frac{3}{10} + \frac{3}{10} + \frac{2}{10}$

   (D) $\frac{4}{10} + \frac{2}{10} + \frac{1}{10}$

5. Lance wants to find the total length of 3 boards. He uses the expression $3\frac{1}{2} + \left(2 + 4\frac{1}{2}\right)$. How can Lance rewrite the expression using both the Associative and Commutative Properties of Addition? (Lesson 7.9)

   (A) $5 + 4\frac{1}{2}$

   (B) $\left(3\frac{1}{2} + 2\right) + 4\frac{1}{2}$

   (C) $2 + \left(3\frac{1}{2} + 4\frac{1}{2}\right)$

   (D) $3\frac{1}{2} + \left(4\frac{1}{2} + 2\right)$

6. Which of the following statements is true? (Lesson 6.6)

   (A) $\frac{5}{8} > \frac{9}{10}$

   (B) $\frac{5}{12} > \frac{1}{3}$

   (C) $\frac{3}{6} > \frac{4}{5}$

   (D) $\frac{1}{2} > \frac{3}{4}$

Multiply a Fraction or Mixed Number by a Whole Number

Multiply. Write the product as a mixed number.

1. $5 \times \dfrac{3}{10} = $ $1\dfrac{5}{10}$ _____

2. $3 \times \dfrac{3}{5} = $ _____

3. $5 \times \dfrac{3}{4} = $ _____

4. $4 \times 1\dfrac{1}{5} = $ _____

5. $2 \times 2\dfrac{1}{3} = $ _____

6. $5 \times 1\dfrac{1}{6} = $ _____

7. $2 \times 2\dfrac{7}{8} = $ _____

8. $7 \times 1\dfrac{3}{4} = $ _____

9. $8 \times 1\dfrac{3}{5} = $ _____

Problem Solving  REAL WORLD

10. Brielle exercises for $\dfrac{3}{4}$ hour each day for 6 days in a row. Altogether, how many hours does she exercise during the 6 days?

11. A recipe for quinoa calls for $2\dfrac{2}{3}$ cups of milk. Conner wants to make 4 batches of quinoa. How much milk does he need?

_____

_____

Lesson Check

1. A mother is $1\frac{3}{4}$ times as tall as her son. Her son is 3 feet tall. How tall is the mother?

   (A) $4\frac{3}{4}$ feet

   (B) $5\frac{1}{4}$ feet

   (C) $5\frac{1}{2}$ feet

   (D) $5\frac{3}{4}$ feet

2. The cheerleaders are making a banner that is 8 feet wide. The length of the banner is $1\frac{1}{3}$ times the width of the banner. How long is the banner?

   (A) $8\frac{1}{3}$ feet

   (B) $8\frac{3}{8}$ feet

   (C) $10\frac{1}{3}$ feet

   (D) $10\frac{2}{3}$ feet

Spiral Review

3. Karleigh walks $\frac{5}{8}$ mile to school every day. How far does she walk to school in 5 days? (Lesson 8.3)

   (A) $\frac{5}{40}$ mile

   (B) $\frac{25}{40}$ mile

   (C) $\frac{10}{8}$ miles

   (D) $\frac{25}{8}$ miles

4. Which number is a multiple of $\frac{4}{5}$? (Lesson 8.2)

   (A) $\frac{8}{10}$

   (B) $\frac{12}{15}$

   (C) $\frac{16}{20}$

   (D) $\frac{12}{5}$

5. Jo cut a key lime pie into 8 equal-size slices. The next day, $\frac{7}{8}$ of the pie is left. Jo puts each slice on its own plate. How many plates does she need? (Lesson 8.1)

   (A) 5

   (B) 6

   (C) 7

   (D) 8

6. Over the weekend, Ed spent $1\frac{1}{4}$ hours doing his math homework and $1\frac{3}{4}$ hours doing his science project. Altogether, how much time did Ed spend doing homework over the weekend? (Lesson 7.7)

   (A) 3 hours

   (B) $2\frac{3}{4}$ hours

   (C) $2\frac{1}{2}$ hours

   (D) 2 hours

Name _____

Problem Solving • Comparison Problems with Fractions

Read each problem and solve.

1. A shrub is $1\frac{2}{3}$ feet tall. A small tree is 3 times as tall as the shrub. How tall is the tree?

   t is the height of the tree, in feet.

   $t = 3 \times 1\frac{2}{3}$

   $t = 3 \times \frac{5}{3}$

   $t = \frac{15}{3}$

   $t = 5$

   So, the tree is 5 feet tall.

   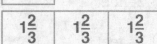

   shrub $\boxed{1\frac{2}{3}}$

   tree $\boxed{1\frac{2}{3} \mid 1\frac{2}{3} \mid 1\frac{2}{3}}$

   _____**5 feet**_____

2. You run $1\frac{3}{4}$ miles each day. Your friend runs 4 times as far as you do. How far does your friend run each day?

   _____

3. At the grocery store, Ayla buys $1\frac{1}{3}$ pounds of ground turkey. Tasha buys 2 times as much ground turkey as Ayla. How much ground turkey does Tasha buy?

   _____

4. When Nathan's mother drives him to school, it takes $\frac{1}{5}$ hour. When Nathan walks to school, it takes him 4 times as long to get to school. How long does it take Nathan to walk to school?

   _____

Lesson Check

1. A Wilson's Storm Petrel is a small bird with a wingspan of $1\frac{1}{3}$ feet. A California Condor is a larger bird with a wingspan almost 7 times as wide as the wingspan of the petrel. About how wide is the wingspan of the California Condor?

   (A) $\frac{4}{21}$ foot

   (B) $2\frac{1}{3}$ feet

   (C) $7\frac{1}{3}$ feet

   (D) $9\frac{1}{3}$ feet

2. The walking distance from the Empire State Building in New York City to Times Square is about $\frac{9}{10}$ mile. The walking distance from the Empire State Building to Sue's hotel is about 8 times as far. About how far is Sue's hotel from the Empire State Building?

   (A) $\frac{9}{80}$ mile

   (B) $\frac{72}{80}$ mile

   (C) $1\frac{7}{10}$ miles

   (D) $7\frac{2}{10}$ miles

Spiral Review

3. Which of the following expressions is NOT equal to $3 \times 2\frac{1}{4}$? (Lesson 8.4)

   (A) $3 \times \frac{9}{4}$

   (B) $(3 \times 2) + \left(3 \times \frac{1}{4}\right)$

   (C) $6\frac{3}{4}$

   (D) $3 \times 2 + \frac{1}{4}$

4. At a bake sale, Ron sells $\frac{7}{8}$ of an apple pie and $\frac{5}{8}$ of a cherry pie. Altogether, how much pie does he sell at the bake sale? (Lesson 7.5)

   (A) $\frac{2}{8}$

   (B) $\frac{12}{16}$

   (C) $\frac{12}{8}$

   (D) $\frac{35}{8}$

5. On a ruler, which measurement is between $\frac{3}{16}$ inch and $\frac{7}{8}$ inch? (Lesson 6.8)

   (A) $\frac{1}{16}$ inch

   (C) $\frac{11}{16}$ inch

   (B) $\frac{1}{8}$ inch

   (D) $\frac{15}{16}$ inch

6. Which of the following numbers is composite? (Lesson 5.5)

   (A) 4

   (C) 2

   (B) 3

   (D) 1

Chapter 8 Extra Practice

Lesson 8.1

Write the fraction as a product of a whole number and a unit fraction.

1. $\frac{5}{6}$ = _____

2. $\frac{7}{8}$ = _____

3. $\frac{3}{5}$ = _____

List the next four multiples of the unit fraction.

4. $\frac{1}{2}$, ____, ____, ____, ____

5. $\frac{1}{6}$, ____, ____, ____, ____

Lesson 8.2

List the next four multiples of the fraction.

1. $\frac{3}{10}$, ____, ____, ____, ____

2. $\frac{7}{12}$, ____, ____, ____, ____

Write the product as the product of a whole number and a unit fraction.

3.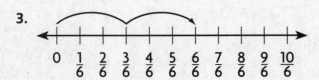

$2 \times \frac{3}{6}$ = _____

4.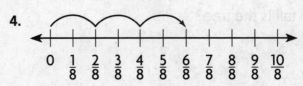

$3 \times \frac{2}{8}$ = _____

Lesson 8.3

Multiply.

1. $3 \times \dfrac{7}{10} =$ _____

2. $5 \times \dfrac{4}{8} =$ _____

3. $4 \times \dfrac{6}{12} =$ _____

4. $2 \times \dfrac{3}{4} =$ _____

5. $6 \times \dfrac{3}{5} =$ _____

6. $7 \times \dfrac{2}{10} =$ _____

Lesson 8.4

Multiply. Write the product as a mixed number.

1. $4 \times \dfrac{8}{10} =$ _____

2. $3 \times \dfrac{5}{6} =$ _____

3. $2 \times 3\dfrac{1}{3} =$ _____

4. $4 \times 2\dfrac{2}{5} =$ _____

5. $5 \times 1\dfrac{7}{8} =$ _____

6. $3 \times 3\dfrac{3}{4} =$ _____

Lesson 8.5

1. A shrub in Pam's back yard is about $1\dfrac{3}{8}$ feet tall. A small tree in her back yard is 7 times as tall as the shrub. About how tall is the tree?

   _____

2. A puppy weighs $\dfrac{9}{10}$ pound. Its mother weighs 8 times as much. How much does the mother weigh?

   _____

School-Home Letter

Dear Family,

During the next few weeks, our math class will relate both fractions and money to place value and will learn how to rename fractions as decimals. We will also add fractional parts of 10 and 100 and compare decimals through hundredths.

You can expect to see homework that provides practice with naming decimals in different ways, including renaming as fractions.

Here is a sample of how your child will be taught to write a decimal as a fraction.

Vocabulary

decimal A number with one or more digits to the right of the decimal point

decimal point A symbol used to separate dollars from cents in money amounts and to separate the ones and tenths places in a decimal

equivalent decimals Two or more decimals that name the same amount

hundredth One of one hundred equal parts

tenth One of ten equal parts

🔒 MODEL Write Hundredths as a Fraction

This is how we will use place value to help write a decimal as a fraction.

Ones	.	Tenths	Hundredths
0	.	6	4

↑
decimal point

Think: 0.64 is the same as 6 tenths and 4 hundredths, or 64 hundredths.

So, $0.64 = \frac{64}{100}$.

Tips

A place-value chart can be used to help visually organize numbers in relation to the decimal place. The chart can be used to pair the numbers with words, and may enable a smooth transition between standard form, word form, and the decimal or fraction.

Activity

Use the relationship between dollars and cents and work together to express the value of a penny, nickel, dime, and quarter as a decimal and as a fraction of a dollar. Then make small groups of coins and help your child write the value of each group as a decimal and as a fraction.

Carta
para la casa

Vocabulario

decimal Un número con uno o más dígitos a la derecha del punto decimal

punto decimal Un símbolo usado para separar dólares de centavos en cantidades de dinero y para separar el lugar de las unidades y los décimos en decimales

decimales equivalentes Dos o más decimales que nombran la misma cantidad

centésimo Una de cien partes iguales

décimo Una de diez partes iguales

Querida familia,

Durante las próximas semanas, en la clase de matemáticas relacionaremos tanto las fracciones como el dinero con el valor posicional y aprenderemos a convertir fracciones en decimales. También sumaremos partes fraccionales de 10 y de 100 y compararemos decimales hasta los centésimos.

Llevaré a la casa tareas para practicar la expresión de decimales de diferentes maneras, incluso la conversión en fracciones.

Este es un ejemplo de la manera como aprenderemos a escribir un decimal como una fracción.

🔒 MODELO Escribir centésimos como una fracción

Así es como usaremos el valor posicional para escribir un decimal como una fracción

Unidades	.	Décimos	Centésimos
0	.	6	4

↑
punto decimal

Piensa: 0.64 es lo mismo que 6 décimos y cuatro centésimos, o 64 centésimos.

Por tanto, $0.64 = \frac{64}{100}$.

Pistas

Una tabla de valor posicional se puede usar para ayudar a organizar visualmente números en relación con el lugar decimal. La tabla puede usarse para emparejar números con palabras y para facilitar la transición del uso de la forma normal a la forma en palabras y a la fracción decimal.

Actividad

Usen la relación entre dólares y centavos y trabajen juntos para expresar el valor de una moneda de uno, de cinco, de diez y de veinticinco centavos en forma decimal y como una fracción de dólar. Luego hagan pequeños grupos de monedas y ayude a su hijo/a a escribir el valor de cada grupo en forma decimal y como fracción.

Name _____

Relate Tenths and Decimals

Write the fraction or mixed number and the decimal shown by the model.

1. 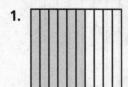 Think: The model is divided into 10 equal parts. Each part represents one tenth.

2.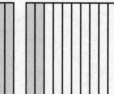

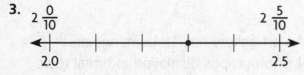 $\frac{6}{10}$; 0.6

_____

_____

3.
2 $\frac{0}{10}$ 2 $\frac{5}{10}$

2.0   2.5

_____

4.
4 $\frac{0}{10}$ 4 $\frac{5}{10}$ 4 $\frac{10}{10}$

4.0   4.5   5.0

_____

Write the fraction or mixed number as a decimal.

5. $\frac{4}{10}$ 6. $3\frac{1}{10}$ 7. $\frac{7}{10}$ 8. $6\frac{5}{10}$ 9. $\frac{9}{10}$

_____   _____   _____   _____   _____

Problem Solving 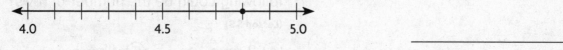 REAL WORLD

10. There are 10 sports balls in the equipment closet. Three are kickballs. Write the portion of the balls that are kickballs as a fraction, as a decimal, and in word form.

11. Peyton has 2 pizzas. Each pizza is cut into 10 equal slices. She and her friends eat 14 slices. What part of the pizzas did they eat? Write your answer as a decimal.

_____   _____

Lesson Check

1. Valerie has 10 CDs in her music case. Seven of the CDs are pop music CDs. What is this amount written as a decimal?

(A) 70.0

(B) 7.0

(C) 0.7

(D) 0.07

2. Which decimal amount is modeled below?

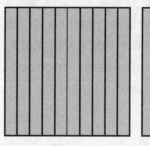

(A) 140.0   (C) 1.4

(B) 14.0   (D) 0.14

Spiral Review

3. Which number is a factor of 13? (Lesson 5.1)

(A) 1

(B) 3

(C) 4

(D) 7

4. An art gallery has 18 paintings and 4 photographs displayed in equal rows on a wall, with the same number of each type of art in each row. Which of the following could be the number of rows? (Lesson 5.3)

(A) 2 rows   (C) 4 rows

(B) 3 rows   (D) 6 rows

5. How do you write the mixed number shown as a fraction greater than 1? (Lesson 7.6)

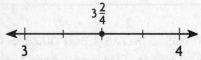

(A) $\frac{32}{5}$ (C) $\frac{6}{4}$

(B) $\frac{14}{4}$ (D) $\frac{4}{4}$

6. Which of the following models has an amount shaded that is equivalent to the fraction $\frac{1}{5}$? (Lesson 6.1)

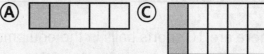

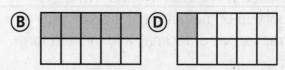

Name _____

Relate Hundredths and Decimals

Write the fraction or mixed number and the decimal shown by the model.

1. 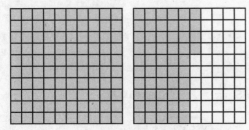 Think: The whole is divided into one hundred equal parts, so each part is one hundredth.

$\frac{77}{100}$; 0.77

2.

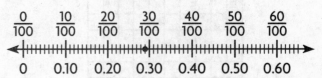

_____

3.

[model: two hundredths grids]

_____

4.

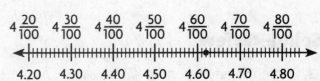

_____

Write the fraction or mixed number as a decimal.

5. $\frac{37}{100}$ **6.** $8\frac{11}{100}$ **7.** $\frac{98}{100}$ **8.** $25\frac{50}{100}$ **9.** $\frac{6}{100}$

_____   _____   _____   _____   _____

Problem Solving REAL WORLD

10. There are 100 pennies in a dollar. What fraction of a dollar is 61 pennies? Write it as a fraction, as a decimal, and in word form.

_____

_____

11. Kylee has collected 100 souvenir thimbles from different places she has visited with her family. Twenty of the thimbles are carved from wood. Write the fraction of thimbles that are wooden as a decimal.

_____

_____

Lesson Check

1. Which decimal represents the shaded section of the model below?

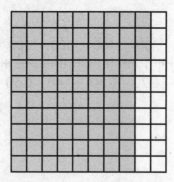

   Ⓐ 830.0   Ⓒ 8.30

   Ⓑ 83.0   Ⓓ 0.83

2. There were 100 questions on the unit test. Alondra answered 97 of the questions correctly. What decimal represents the fraction of questions Alondra answered correctly?

   Ⓐ   0.97

   Ⓑ   9.70

   Ⓒ 90.70

   Ⓓ 970.0

Spiral Review

3. Which is equivalent to $\frac{7}{8}$? (Lesson 7.2)

   Ⓐ $\frac{5}{8} + \frac{3}{8}$

   Ⓑ $\frac{4}{8} + \frac{1}{8} + \frac{1}{8}$

   Ⓒ $\frac{3}{8} + \frac{2}{8} + \frac{2}{8}$

   Ⓓ $\frac{2}{8} + \frac{2}{8} + \frac{1}{8} + \frac{1}{8}$

4. What is $\frac{9}{10} - \frac{6}{10}$? (Lesson 7.4)

   Ⓐ $\frac{1}{10}$ Ⓒ $\frac{4}{10}$

   Ⓑ $\frac{3}{10}$ Ⓓ $\frac{6}{10}$

5. Misha used $\frac{1}{4}$ of a carton of 12 eggs to make an omelet. How many eggs did she use? (Lesson 8.4)

   Ⓐ 2

   Ⓑ 3

   Ⓒ 4

   Ⓓ 6

6. Kurt used the rule *add* 4, *subtract* 1 to generate a pattern. The first term in his pattern is 5. Which number could be in Kurt's pattern? (Lesson 5.6)

   Ⓐ   4

   Ⓑ   6

   Ⓒ 10

   Ⓓ 14

Equivalent Fractions and Decimals

Write the number as hundredths in fraction form and decimal form.

1. $\dfrac{5}{10}$

$\dfrac{5}{10} = \dfrac{5 \times 10}{10 \times 10} = \dfrac{50}{100}$

Think: 5 tenths is the same as 5 tenths and 0 hundredths. Write 0.50.

$$\dfrac{50}{100}; \ 0.50$$

2. $\dfrac{9}{10}$

3. 0.2

4. 0.8

Write the number as tenths in fraction form and decimal form.

5. $\dfrac{40}{100}$

6. $\dfrac{10}{100}$

7. 0.60

Problem Solving REAL WORLD

8. Billy walks $\dfrac{6}{10}$ mile to school each day. Write $\dfrac{6}{10}$ as hundredths in fraction form and in decimal form.

9. Four states have names that begin with the letter A. This represents 0.08 of all the states. Write 0.08 as a fraction.

Lesson Check

1. The fourth-grade students at Harvest School make up 0.3 of all students at the school. Which fraction is equivalent to 0.3?

   (A) $\frac{3}{10}$

   (B) $\frac{30}{10}$

   (C) $\frac{3}{100}$

   (D) $\frac{33}{100}$

2. Kyle and his brother have a marble set. Of the marbles, 12 are blue. This represents $\frac{50}{100}$ of all the marbles. Which decimal is equivalent to $\frac{50}{100}$?

   (A)   50

   (B)   5.0

   (C)   0.50

   (D) 5,000

Spiral Review

3. Jesse won his race by $3\frac{45}{100}$ seconds. What is this number written as a decimal? (Lesson 9.2)

   (A)   0.345

   (B)   3.45

   (C)   34.5

   (D) 345

4. Marge cut 16 pieces of tape for mounting pictures on poster board. Each piece of tape was $\frac{3}{8}$ inch long. How much tape did Marge use? (Lesson 8.4)

   (A) 2 inches

   (B) 4 inches

   (C) 5 inches

   (D) 6 inches

5. Of Katie's pattern blocks, $\frac{9}{12}$ are triangles. What is $\frac{9}{12}$ in simplest form? (Lesson 6.3)

   (A) $\frac{1}{4}$

   (B) $\frac{2}{3}$

   (C) $\frac{3}{4}$

   (D) $\frac{9}{12}$

6. A number pattern has 75 as its first term. The rule for the pattern is *subtract* 6. What is the sixth term? (Lesson 5.6)

   (A) 39

   (B) 45

   (C) 51

   (D) 69

Name _____

Relate Fractions, Decimals, and Money

Write the total money amount. Then write the amount as a fraction or a mixed number and as a decimal in terms of dollars.

1.

$0.18; $\frac{18}{100}$; 0.18

2.

_____

Write as a money amount and as a decimal in terms of dollars.

3. $\frac{25}{100}$ 4. $\frac{79}{100}$ 5. $\frac{31}{100}$ 6. $\frac{8}{100}$ 7. $\frac{42}{100}$

_____   _____   _____   _____   _____

Write the money amount as a fraction in terms of dollars.

8. $0.87   9. $0.03   10. $0.66   11. $0.95   12. $1.00

_____   _____   _____   _____   _____

Write the total money amount. Then write the amount as a fraction and as a decimal in terms of dollars.

13. 2 quarters 2 dimes   14. 3 dimes 4 pennies   15. 8 nickels 12 pennies

_____   _____   _____

Problem Solving REAL WORLD

16. Kate has 1 dime, 4 nickels, and 8 pennies. Write Kate's total amount as a fraction in terms of a dollar.

17. Nolan says he has $\frac{75}{100}$ of a dollar. If he only has 3 coins, what are the coins?

_____   _____

Lesson Check

1. Which of the following names the total money amount shown as a fraction in terms of a dollar?

  Ⓐ $\frac{43}{1}$ Ⓒ $\frac{43}{57}$

  Ⓑ $\frac{43}{10}$ Ⓓ $\frac{43}{100}$

2. Crystal has $\frac{81}{100}$ of a dollar. Which of the following could be the coins Crystal has?

  Ⓐ 3 quarters, 1 dime, 1 penny

  Ⓑ 2 quarters, 6 nickels, 1 penny

  Ⓒ 2 quarters, 21 pennies

  Ⓓ 1 quarter, 4 dimes, 1 nickel, 1 penny

Spiral Review

3. Joel gives $\frac{1}{3}$ of his baseball cards to his sister. Which fraction is equivalent to $\frac{1}{3}$? **(Lesson 6.2)**

  Ⓐ $\frac{3}{5}$ Ⓒ $\frac{8}{9}$

  Ⓑ $\frac{2}{6}$ Ⓓ $\frac{4}{10}$

4. Penelope bakes pretzels. She salts $\frac{3}{8}$ of the pretzels. Which fraction is equivalent to $\frac{3}{8}$? **(Lesson 6.2)**

  Ⓐ $\frac{9}{24}$ Ⓒ $\frac{3}{16}$

  Ⓑ $\frac{15}{20}$ Ⓓ $\frac{1}{5}$

5. Which decimal is shown by the model? **(Lesson 9.2)**

  Ⓐ 10.0   Ⓒ 0.1

  Ⓑ 1.0   Ⓓ 0.01

6. Mr. Guzman has 100 cows on his dairy farm. Of the cows, 57 are Holstein. What decimal represents the portion of cows that are Holstein? **(Lesson 9.2)**

  Ⓐ 0.43

  Ⓑ 0.57

  Ⓒ 5.7

  Ⓓ 57.0

Name _____

Problem Solving • Money

Use the *act it out* strategy to solve.

1. Carl wants to buy a bicycle bell that costs
   $4.50. Carl has saved $2.75 so far. How much
   more money does he need to buy the bell?

   Use 4 $1 bills and 2 quarters to model $4.50.
   Remove bills and coins that have a value of
   $2.75. First, remove 2 $1 bills and 2 quarters.

   Next, exchange one $1 bill for 4 quarters and
   remove 1 quarter.

   Count the amount that is left.
   So, Carl needs to save $1.75 more.

   _____ **$1.75**

2. Together, Xavier, Yolanda, and Zachary have
   $4.44. If each person has the same amount,
   how much money does each person have?

   _____

3. Marcus, Nan, and Olive each have $1.65
   in their pockets. They decide to combine
   the money. How much money do they have
   altogether?

   _____

4. Jessie saves $6 each week. In how many
   weeks will she have saved at least $50?

   _____

5. Becca has $12 more than Cece. Dave has
   $3 less than Cece. Cece has $10. How much
   money do they have altogether?

   _____

Lesson Check

1. Four friends earned $5.20 for washing a car. They shared the money equally. How much did each friend get?

   Ⓐ $1.05

   Ⓑ $1.30

   Ⓒ $1.60

   Ⓓ $20.80

2. Which represents the value of one $1 bill and 5 quarters?

   Ⓐ $1.05

   Ⓑ $1.25

   Ⓒ $1.50

   Ⓓ $2.25

Spiral Review

3. Bethany has 9 pennies. What fraction of a dollar is this? (Lesson 9.4)

   Ⓐ $\frac{9}{100}$

   Ⓑ $\frac{9}{10}$

   Ⓒ $\frac{90}{100}$

   Ⓓ $\frac{99}{100}$

4. Michael made $\frac{9}{12}$ of his free throws at practice. What is $\frac{9}{12}$ in simplest form?
   (Lesson 6.3)

   Ⓐ $\frac{1}{4}$

   Ⓑ $\frac{3}{9}$

   Ⓒ $\frac{1}{2}$

   Ⓓ $\frac{3}{4}$

5. I am a prime number between 30 and 40. Which number could I be? (Lesson 5.5)

   Ⓐ 31

   Ⓑ 33

   Ⓒ 36

   Ⓓ 39

6. Georgette is using the benchmark $\frac{1}{2}$ to compare fractions. Which statement is correct? (Lesson 6.6)

   Ⓐ $\frac{3}{8} > \frac{1}{2}$

   Ⓑ $\frac{2}{5} < \frac{1}{2}$

   Ⓒ $\frac{7}{12} < \frac{1}{2}$

   Ⓓ $\frac{9}{10} = \frac{1}{2}$

Name _____

Add Fractional Parts of 10 and 100

Find the sum.

1. $\frac{2}{10} + \frac{43}{100}$

   $\frac{20}{100} + \frac{43}{100} = \frac{63}{100}$

   $$\frac{63}{100}$$

Think: Write $\frac{2}{10}$ as a fraction with a denominator of 100:

$\frac{2 \times 10}{10 \times 10} = \frac{20}{100}$

2. $\frac{17}{100} + \frac{6}{10}$

3. $\frac{9}{100} + \frac{4}{10}$

4. $\frac{7}{10} + \frac{23}{100}$

5. $\$0.48 + \$0.30$

6. $\$0.25 + \$0.34$

7. $\$0.66 + \$0.06$

Problem Solving REAL WORLD

8. Ned's frog jumped $\frac{38}{100}$ meter. Then his frog jumped $\frac{4}{10}$ meter. How far did Ned's frog jump in all?

9. Keiko walks $\frac{5}{10}$ kilometer from school to the park. Then she walks $\frac{19}{100}$ kilometer from the park to her home. How far does Keiko walk in all?

Lesson Check

1. In a fish tank, $\frac{2}{10}$ of the fish were orange and $\frac{5}{100}$ of the fish were striped. What fraction of the fish were orange or striped?

   (A) $\frac{7}{10}$

   (B) $\frac{52}{100}$

   (C) $\frac{25}{100}$

   (D) $\frac{7}{100}$

2. Greg spends $0.45 on an eraser and $0.30 on a pen. How much money does Greg spend in all?

   (A) $3.45

   (B) $0.75

   (C) $0.48

   (D) $0.15

Spiral Review

3. Phillip saves $8 each month. How many months will it take him to save at least $60? **(Lesson 9.5)**

   (A) 6 months

   (B) 7 months

   (C) 8 months

   (D) 9 months

4. Ursula and Yi share a submarine sandwich. Ursula eats $\frac{2}{8}$ of the sandwich. Yi eats $\frac{3}{8}$ of the sandwich. How much of the sandwich do the two friends eat?

   (Lesson 7.5)

   (A) $\frac{1}{8}$

   (B) $\frac{4}{8}$

   (C) $\frac{5}{8}$

   (D) $\frac{6}{8}$

5. A carpenter has a board that is 8 feet long. He cuts off two pieces. One piece is $3\frac{1}{2}$ feet long and the other is $2\frac{1}{3}$ feet long. How much of the board is left? **(Lesson 7.10)**

   (A) $2\frac{1}{6}$ feet

   (B) $2\frac{5}{6}$ feet

   (C) $3\frac{1}{6}$ feet

   (D) $3\frac{5}{6}$ feet

6. Jeff drinks $\frac{2}{3}$ of a glass of juice. Which fraction is equivalent to $\frac{2}{3}$? **(Lesson 6.2)**

   (A) $\frac{1}{3}$

   (B) $\frac{3}{2}$

   (C) $\frac{3}{6}$

   (D) $\frac{8}{12}$

Name _____

Compare Decimals

Compare. Write <, >, or =.

1. 0.35 $\lt$ 0.53

Think: 3 tenths is less
than 5 tenths.
So, 0.35 < 0.53

2. 0.6 ◯ 0.60

3. 0.24 ◯ 0.31

4. 0.94 ◯ 0.9 | **5.** 0.3 ◯ 0.32 | **6.** 0.45 ◯ 0.28 | **7.** 0.39 ◯ 0.93

Use the number line to compare. Write *true* or *false*.

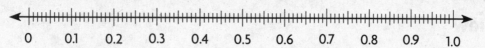

8. 0.8 > 0.78 | **9.** 0.4 > 0.84 | **10.** 0.7 < 0.70 | **11.** 0.4 > 0.04

_____ | _____ | _____ | _____

Compare. Write *true* or *false*.

12. 0.09 > 0.1 | **13.** 0.24 = 0.42 | **14.** 0.17 < 0.32 | **15.** 0.85 > 0.82

_____ | _____ | _____ | _____

Problem Solving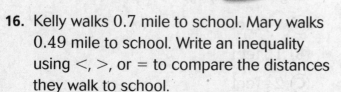

16. Kelly walks 0.7 mile to school. Mary walks 0.49 mile to school. Write an inequality using <, >, or = to compare the distances they walk to school.

17. Tyrone shades two decimal grids. He shades 0.03 of the squares on one grid blue. He shades 0.3 of another grid red. Which grid has the greater part shaded?

_____ | _____

Lesson Check

1. Bob, Cal, and Pete each made a stack of baseball cards. Bob's stack was 0.2 meter high. Cal's stack was 0.24 meter high. Pete's stack was 0.18 meter high. Which statement is true?

   (A) 0.2 > 0.24

   (B) 0.24 > 0.18

   (C) 0.18 > 0.2

   (D) 0.24 = 0.2

2. Three classmates spent money at the school supplies store. Mark spent 0.5 dollar, Andre spent 0.45 dollar, and Raquel spent 0.52 dollar. Which statement is true?

   (A) 0.45 > 0.5

   (B) 0.52 < 0.45

   (C) 0.5 = 0.52

   (D) 0.45 < 0.5

Spiral Review

3. Pedro has $0.35 in his pocket. Alice has $0.40 in her pocket. How much money do Pedro and Alice have in their pockets altogether? (Lesson 9.6)

   (A) $0.05

   (B) $0.39

   (C) $0.75

   (D) $0.79

4. The measure 62 centimeters is equivalent to $\frac{62}{100}$ meter. What is this measure written as a decimal? (Lesson 9.3)

   (A) 62.0 meters

   (B) 6.2 meters

   (C) 0.62 meter

   (D) 0.6 meter

5. Joel has 24 sports trophies. Of the trophies, $\frac{1}{8}$ are soccer trophies. How many soccer trophies does Joel have?

   (Lesson 8.4)

   (A) 2

   (B) 3

   (C) 4

   (D) 6

6. Molly's jump rope is $6\frac{1}{3}$ feet long. Gail's jump rope is $4\frac{2}{3}$ feet long. How much longer is Molly's jump rope? (Lesson 7.8)

   (A) $1\frac{1}{3}$ feet

   (B) $1\frac{2}{3}$ feet

   (C) $2\frac{1}{3}$ feet

   (D) $2\frac{2}{3}$ feet

Name _____

Chapter 9 Extra Practice

Lessons 9.1 - 9.2

Write the fraction or mixed number and the decimal shown by the model.

1.

2.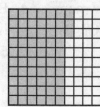

3.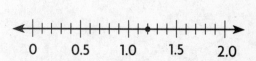

_____   _____   _____

Lesson 9.3

Write the number as hundredths in fraction form and decimal form.

1. $\frac{8}{10}$

2. 0.1

3. $\frac{3}{10}$

_____   _____   _____

Write the number as tenths in fraction form and decimal form.

4. $\frac{60}{100}$

5. $\frac{70}{100}$

6. 0.20

_____   _____   _____

Lesson 9.4

Write as a money amount and as a decimal in terms of dollars.

1. $\frac{30}{100}$

2. $\frac{91}{100}$

3. $\frac{5}{100}$

_____   _____   _____

Write the total money amount. Then write the amount as a fraction and as a decimal in terms of dollars.

4. 4 dimes, 9 pennies

5. 3 quarters, 1 dime

6. 7 nickels, 2 pennies

_____   _____   _____

Lesson 9.5

1. Camila, Jocelyn, and Audrey each earned $2.55. How much did the three girls earn altogether?

   _____

2. Elijah, Xavier, and Adrian earned a total of $8.34. The boys shared the earnings equally. How much did each boy get?

   _____

3. Anthony saves $7 each week. In how many weeks will he have saved at least $40?

   _____

4. Brianna has $2 less than Victoria. Victoria has $11 more than Damian. Damian has $6. How much money do they have in all?

   _____

Lesson 9.6

Find the sum.

1. $\frac{6}{10} + \frac{39}{100}$

   _____

2. $\frac{14}{100} + \frac{8}{10}$

   _____

3. $\frac{4}{10} + \frac{18}{100}$

   _____

4. $\frac{5}{10} + \frac{16}{100}$

   _____

5. $0.43 + $0.20

   _____

6. $0.07 + $0.35

   _____

7. $0.80 + $0.15

   _____

8. $0.52 + $0.28

   _____

Lesson 9.7

Compare. Write $<$, $>$, or $=$.

1. 0.3 ◯ 0.39

2. 0.9 ◯ 0.90

3. 0.54 ◯ 0.45

4. 0.04 ◯ 0.06

5. 0.7 ◯ 0.70

6. 0.36 ◯ 0.51

7. 0.8 ◯ 0.67

8. 0.63 ◯ 0.48

Compare. Write *true* or *false*.

9. 0.32 > 0.23

10. 0.86 = 0.9

11. 0.68 < 0.83

12. 0.97 > 0.94

_____

Dear Family,

Throughout the next few weeks, our math class will be studying two-dimensional figures. The students will use definitions to identify and describe characteristics of these figures.

You can expect to see homework that includes identifying types of triangles and quadrilaterals.

Here is a sample of how your child will be taught to classify a triangle by its angles.

Vocabulary

acute triangle A triangle with three acute angles

line segment A part of a line that includes two points, called endpoints, and all the points between them

obtuse triangle A triangle with one obtuse angle

ray A part of a line, with one endpoint, that is straight and continues in one direction

right triangle A triangle with one right angle and two acute angles

🔑 MODEL Classify a triangle by the sizes of its angles.

Classify triangle *KLM*.

STEP 1

Determine how many angles are acute.

∠K is ___acute___.

∠L is ___acute___.

∠M is ___acute___.

STEP 2

Determine the correct classification.

A triangle with ___3___ acute angles is

___acute___.

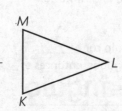

Tips

Angle sizes

Angles are classified by the size of the opening between the rays. A right angle forms a square corner. An acute angle is less than a right angle. An obtuse angle is greater than a right angle and less than a straight angle.

To classify angles in a figure, use the corner of an index card as a right angle and compare.

Activity

Help your child commit most of the classifications of triangles and quadrilaterals to memory. Together, you can make a series of flash cards with the classifications on one side of the card and definitions and/or sketches of examples on the other side of the card.

Carta
para la casa

Vocabulario

triángulo agudo Un triángulo que tiene tres ángulos agudos

segmento de recta Una parte de una línea que incluye dos puntos, llamados extremos, y los puntos que están entre ellos

triángulo obtuso Un triángulo que tiene un ángulo obtuso

rayo Parte de una línea recta, con un extremo y que continúa en una dirección

triángulo rectángulo Un triángulo con un ángulo recto y dos ángulos agudos

Querida familia,

Durante las próximas semanas, en la clase de matemáticas estudiaremos las figuras bidimensionales. Usaremos las definiciones para identificar y describir las características de esas figuras.

Llevaré a la casa tareas con actividades para identificar diferentes tipos de triángulos y cuadriláteros.

Este es un ejemplo de la manera como aprenderemos a clasificar un triángulo por sus ángulos.

🔑 MODELO Clasificar un triángulo por el tamaño de sus lados

Clasifica el triángulo *KLM*.

PASO 1

Identifica cuántos ángulos son agudos.

$\angle K$ es ___**agudo**___.

$\angle L$ es ___**agudo**___.

$\angle M$ es ___**agudo**___.

PASO 2

Determina la clasificación correcta.

Un triángulo con ___**3**___ ángulos agudos, entonces es

___**acutángulo**___.

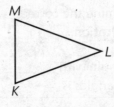

Pistas

Tipos de ángulos

Los ángulos se clasifican según el tamaño de la abertura entre sus rayos. Un ángulo recto forma una esquina recta. Un ángulo agudo mide menos que un ángulo recto. Un ángulo obtuso mide más que un ángulo recto y menos que un ángulo llano.

Para clasificar los ángulos de una figura, usa la esquina de una tarjeta como modelo de ángulo recto y compara.

Actividad

Anime a su hijo a memorizar las clasificaciones de los triángulos y los cuadriláteros. Puede hacer tarjetas nemotécnicas con las clasificaciones en un lado y las definiciones y/o ejemplos visuales en el otro lado de cada tarjeta.

Name _____

Lines, Rays, and Angles

Draw and label an example of the figure.

1. obtuse ∠ABC

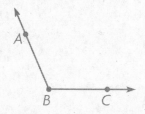

Think: An obtuse angle is greater than a right angle. The middle letter, B, names the vertex of the angle.

2. $\overrightarrow{GH}$

3. acute ∠JKL

4. $\overline{BC}$

Use the figure for 5–8.

5. Name a line segment.

_____

6. Name a right angle.

_____

7. Name an obtuse angle.

_____

8. Name a ray.

_____

Problem Solving REAL WORLD

Use the figure at the right for 9–11.

9. Classify ∠AFD. _____

10. Classify ∠CFE. _____

11. Name two acute angles.

_____

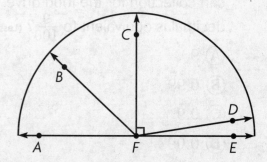

Lesson Check

1. The hands of a clock show the time 12:25.

   Which best describes the angle between the hands of the clock?

   (A) acute   (C) obtuse

   (B) right   (D) straight

2. Which of the following name two different figures?

   (A) $\overline{AB}$ and $\overline{BA}$

   (B) $\overrightarrow{AB}$ and $\overrightarrow{BA}$

   (C) $\overline{AB}$ and $\overrightarrow{BA}$

   (D) $\angle ABC$ and $\angle CBA$

Spiral Review

3. Jan's pencil is 8.5 cm long. Ted's pencil is longer. Which could be the length of Ted's pencil? **(Lesson 9.7)**

   (A) 0.09 cm

   (B) 0.8 cm

   (C) 8.4 cm

   (D) 9.0 cm

4. Kayla buys a shirt for $8.19. She pays with a $10 bill. How much change should she receive? **(Lesson 9.5)**

   (A) $1.81

   (B) $1.89

   (C) $2.19

   (D) $2.81

5. Sasha donated $\frac{9}{100}$ of her class's entire can collection for the food drive. Which decimal is equivalent to $\frac{9}{100}$? **(Lesson 9.2)**

   (A) 9

   (B) 0.99

   (C) 0.9

   (D) 0.09

6. Jose jumped $8\frac{1}{3}$ feet. This was $2\frac{2}{3}$ feet farther than Lila jumped. How far did Lila jump? **(Lesson 7.8)**

   (A) $5\frac{1}{3}$ feet

   (B) $5\frac{2}{3}$ feet

   (C) $6\frac{1}{3}$ feet

   (D) 11 feet

Classify Triangles

Classify each triangle. Write *acute*, *right*, or *obtuse*.

1.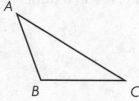

Think: Angles *A* and *C* are both acute.
Angle *B* is obtuse.

_____obtuse_____

2.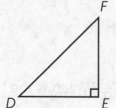

3.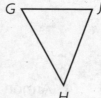

4.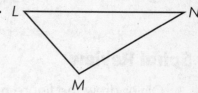

_____

_____

_____

Problem Solving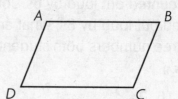

5. Use figure *ABCD* below. Draw a line segment from point *B* to point *D*. Name and classify the triangles formed.

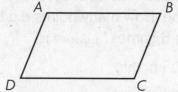

_____

_____

_____

6. Use figure *ABCD* below. Draw a line segment from point *A* to point *C*. Name and classify the triangles formed.

_____

_____

_____

Lesson Check

1. Stephen drew this triangle. How many obtuse angles does the triangle have?

   (A) 0   **(C)** 2

   (B) 1   **(D)** 3

2. Joan was asked to draw a right triangle. How many right angles are in a right triangle?

   (A) 0

   (B) 1

   (C) 2

   (D) 3

Spiral Review

3. Oliver drew the figure below to show light traveling from the sun to Earth. Name the figure he drew. **(Lesson 10.1)**

   (A) segment *SE*   **(C)** line *SE*

   (B) ray *SE*   **(D)** ray *ES*

4. Armon added $\frac{1}{10}$ and $\frac{8}{100}$. Which is the correct sum? **(Lesson 9.6)**

   (A) $\frac{18}{10}$

   (B) $\frac{9}{10}$

   (C) $\frac{9}{100}$

   (D) $\frac{18}{100}$

5. Sam counted out loud by 6s. Jorge counted out loud by 8s. What are the first three numbers both students said?

   (Lesson 5.4)

   (A) 8, 16, 24

   (B) 14, 28, 42

   (C) 24, 48, 72

   (D) 48, 96, 144

6. A basketball team averaged 105 points per game. How many points did the team score in 6 games? **(Lesson 2.10)**

   (A) 605 points

   (B) 630 points

   (C) 900 points

   (D) 6,030 points

Name _____

Parallel Lines and Perpendicular Lines

Use the figure for 1–3.

1. Name a pair of lines that appear to be perpendicular.

   Think: Perpendicular lines form right angles.
   $\overleftrightarrow{AB}$ and $\overleftrightarrow{EF}$ appear to form right angles.

   $\overleftrightarrow{AB}$ **and** $\overleftrightarrow{EF}$

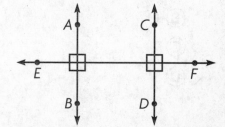

2. Name a pair of lines that appear to be parallel.

   _____

3. Name another pair of lines that appear to be perpendicular.

   _____

Draw and label the figure described.

4. $\overleftrightarrow{MN}$ and $\overleftrightarrow{PQ}$ intersecting at point R

5. $\overleftrightarrow{WX} \parallel \overleftrightarrow{YZ}$

6. $\overleftrightarrow{FH} \perp \overleftrightarrow{JK}$

Problem Solving REAL WORLD

Use the street map for 7–8.

7. Name two streets that intersect but do not appear to be perpendicular.

   _____

   _____

8. Name two streets that appear to be parallel to each other.

   _____

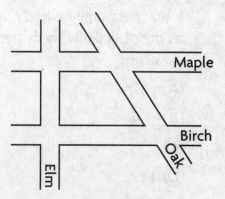

Lesson Check

1. Which capital letter appears to have perpendicular line segments?

   Ⓐ N

   Ⓑ O

   Ⓒ T

   Ⓓ V

2. In the figure, which pair of line segments appear to be parallel?

   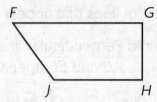

   Ⓐ $\overline{FG}$ and $\overline{GH}$

   Ⓑ $\overline{FJ}$ and $\overline{GH}$

   Ⓒ $\overline{FG}$ and $\overline{JH}$

   Ⓓ $\overline{JH}$ and $\overline{FJ}$

Spiral Review

3. Nolan drew a right triangle. How many acute angles did he draw? (Lesson 10.2)

   Ⓐ 0

   Ⓑ 1

   Ⓒ 2

   Ⓓ 3

4. Mike drank more than half the juice in his glass. What fraction of the juice could Mike have drunk? (Lesson 6.6)

   Ⓐ $\frac{1}{3}$

   Ⓑ $\frac{2}{5}$

   Ⓒ $\frac{3}{6}$

   Ⓓ $\frac{5}{8}$

5. A school principal ordered 1,000 pencils. He gave an equal number to each of 7 teachers until he had given out as many as possible. How many pencils were left?

   (Lesson 4.11)

   Ⓐ 2

   Ⓑ 4

   Ⓒ 6

   Ⓓ 142

6. A carton of juice contains 64 ounces. Ms. Wilson bought 6 cartons of juice. How many ounces of juice did she buy?

   (Lesson 2.10)

   Ⓐ 364 ounces

   Ⓑ 370 ounces

   Ⓒ 384 ounces

   Ⓓ 402 ounces

Name _____

Classify Quadrilaterals

Classify each figure as many ways as possible. Write
quadrilateral, trapezoid, parallelogram, rhombus, rectangle, or square.

1.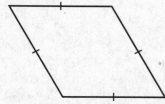

Think: 2 pairs of parallel sides
        4 sides of equal length
        0 right angles

<u>quadrilateral, parallelogram,
rhombus</u>

2.

_____

_____

3.

_____

_____

4.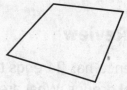

_____

_____

5.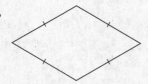

_____

_____

6.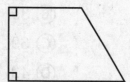

_____

_____

7.

_____

_____

Problem Solving

8. Alan drew a polygon with four sides and four angles. All four sides are equal. None of the angles are right angles. What figure did Alan draw?

_____

_____

9. Teresa drew a quadrilateral with 2 pairs of parallel sides and 4 right angles. What quadrilateral could she have drawn?

_____

_____

Lesson Check

1. Joey is asked to name a quadrilateral that is also a rhombus. What should be his answer?

    (A) square

    (B) rectangle

    (C) parallelogram

    (D) trapezoid

2. Which quadrilateral has exactly one pair of parallel sides?

    (A) square

    (B) rhombus

    (C) parallelogram

    (D) trapezoid

Spiral Review

3. Terrence has 24 eggs to divide into equal groups. What are all the possible numbers of eggs that Terence could put in each group? (Lesson 5.2)

    (A) 1, 2, 3, 4

    (B) 2, 4, 6, 8, 12

    (C) 1, 2, 3, 4, 6, 8, 12, 24

    (D) 24, 48, 72, 96

4. In a line of students, Jenna is number 8. The teacher says that a rule for a number pattern is *add 4*. The first student in line says the first term, 7. What number should Jenna say? (Lesson 5.6)

    (A) 31

    (B) 35

    (C) 39

    (D) 43

5. Lou eats $\frac{6}{8}$ of a pizza. What fraction of the pizza is left over? (Lesson 7.5)

    (A) $\frac{1}{8}$

    (B) $\frac{1}{4}$

    (C) $\frac{1}{2}$

    (D) $\frac{3}{4}$

6. Which capital letter appears to have parallel lines? (Lesson 10.3)

    (A) D

    (B) L

    (C) N

    (D) T

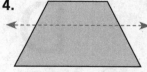

Line Symmetry

Tell if the dashed line appears to be a line of symmetry.
Write *yes* or *no*.

1.

yes

2.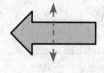

_____

3.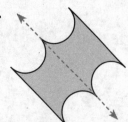

_____

4.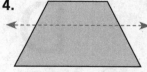

_____

5.

_____

6.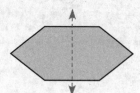

_____

7.

_____

8.

_____

Complete the design by reflecting over the line of symmetry.

9.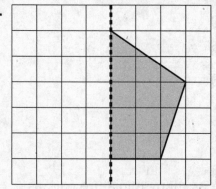

10.

Problem Solving REAL WORLD

11. Kara uses the pattern at the right to make
paper dolls. The dashed line represents a
line of symmetry. A complete doll includes
the reflection of the pattern over the line
of symmetry. Complete the design to show
what one of Kara's paper dolls looks like.

Lesson Check

1. Which best describes the line of symmetry in the letter D?

- (A) horizontal
- (B) vertical
- (C) diagonal
- (D) half turn

2. Which shape has a correctly drawn line of symmetry?

(A)

(C)

(B)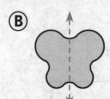

(D)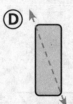

Spiral Review

3. The class has 360 unit cubes in a bag. Johnnie divides the unit cubes equally among 8 groups. How many unit cubes will each group get? **(Lesson 4.11)**

- (A) 40
- (B) 44
- (C) 45
- (D) 48

4. There are 5,280 feet in one mile. How many feet are there in 6 miles? **(Lesson 2.11)**

- (A) 30,680
- (B) 31,260
- (C) 31,608
- (D) 31,680

5. Sue has 4 pieces of wood. The lengths of her pieces of wood are $\frac{1}{3}$ foot, $\frac{2}{5}$ foot, $\frac{3}{10}$ foot, and $\frac{1}{4}$ foot. Which piece of wood is the shortest? **(Lesson 6.7)**

- (A) the $\frac{1}{3}$-foot piece
- (B) the $\frac{2}{5}$-foot piece
- (C) the $\frac{3}{10}$-foot piece
- (D) the $\frac{1}{4}$-foot piece

6. Alice has $\frac{1}{5}$ as many miniature cars as Sylvester has. Sylvester has 35 miniature cars. How many miniature cars does Alice have? **(Lesson 8.5)**

- (A) 7
- (B) 9
- (C) 40
- (D) 175

Name _____

Find and Draw Lines of Symmetry

Tell whether the shape appears to have zero lines, 1 line, or more than 1 line of symmetry. Write *zero, 1,* or *more than 1*.

1.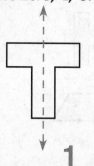

2.

3.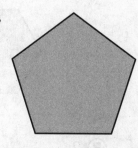

4.

_____1_____   _____   _____   _____

Does the design have line symmetry? Write *yes* or *no*. If your answer is yes, draw all lines of symmetry.

5.

6.

7.

8.

_____   _____   _____   _____

Draw a shape for the statement. Draw the line or lines of symmetry.

9. zero lines of symmetry

10. 1 line of symmetry

11. 2 lines of symmetry

Problem Solving

Use the chart for 12–13.

0 2 3 4
5 6 8 9

12. Which number or numbers appear to have only 1 line of symmetry?

_____

13. Which number or numbers appear to have 2 lines of symmetry?

_____

Lesson Check

1. How many lines of symmetry does this shape appear to have?

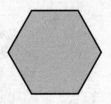

- Ⓐ 0
- Ⓑ 2
- Ⓒ 6
- Ⓓ 10

2. Which of the following shapes appears to have exactly 1 line of symmetry?

Ⓐ

Ⓒ

Ⓑ

Ⓓ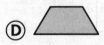

Spiral Review

3. Richard practiced each of 3 piano solos for $\frac{5}{12}$ hour. How long did he practice in all? **(Lesson 8.3)**

- Ⓐ $\frac{2}{3}$ hour
- Ⓑ $1\frac{1}{4}$ hours
- Ⓒ $1\frac{1}{3}$ hours
- Ⓓ $1\frac{5}{12}$ hours

4. Which of the following decimals is equivalent to three and ten hundredths? **(Lesson 9.2)**

- Ⓐ 0.30
- Ⓑ 0.31
- Ⓒ 3.01
- Ⓓ 3.1

5. Lynne used $\frac{3}{8}$ cup of flour and $\frac{1}{3}$ cup of sugar in a recipe. Which number below is a common denominator for $\frac{3}{8}$ and $\frac{1}{3}$?

(Lesson 6.4)

- Ⓐ 8
- Ⓑ 12
- Ⓒ 16
- Ⓓ 24

6. Kevin draws a figure that has four sides. All sides have the same length. His figure has no right angles. What figure does Kevin draw? **(Lesson 10.4)**

- Ⓐ square
- Ⓑ trapezoid
- Ⓒ rhombus
- Ⓓ rectangle

Name _____

Problem Solving • Shape Patterns

Solve each problem.

1. Marta is using this pattern to decorate a picture frame.
   Describe the pattern. Draw what might be the next
   three figures in the pattern.

   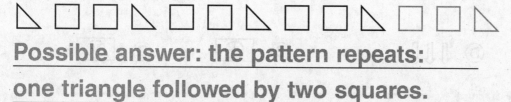

   Possible answer: the pattern repeats:
   one triangle followed by two squares.

2. Describe the pattern. Draw what might be the next
   three figures in the pattern. How many circles are
   in the sixth figure in the pattern?

   _____

   _____

3. Larry stencils this pattern to make a border at the top of
   his bedroom walls. Describe the pattern. Draw what might
   be the missing figure in the pattern.

   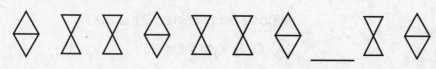

   _____

   _____

Lesson Check

1. What might be the next three figures in this pattern?

⇑⇓⇓⇑⇑⇑⇓⇓⇓⇑⇑⇑⇑⇑⇓⇓⇓⇓

  Ⓐ ⇓⇓⇑   Ⓒ ⇑⇓⇓

  Ⓑ ⇓⇑⇑   Ⓓ ⇓⇓⇓

2. Which might be the missing figure in the following pattern?

  Ⓐ ⊗   Ⓒ ⊕

  Ⓑ ⊕   Ⓓ ⊗

Spiral Review

3. Chad has two pieces of wood. One piece is $\frac{7}{12}$ foot long. The second piece is $\frac{5}{12}$ foot longer than the first piece. How long is the second piece? **(Lesson 7.5)**

  Ⓐ $\frac{2}{12}$ foot

  Ⓑ $\frac{1}{2}$ foot

  Ⓒ $\frac{12}{18}$ foot

  Ⓓ 1 foot

4. Olivia finished a race in 40.64 seconds. Patty finished the race in 40.39 seconds. Miguel finished the race in 41.44 seconds. Chad finished the race in 40.46 seconds. Who finished the race in the least time?

(Lesson 9.7)

  Ⓐ Olivia

  Ⓑ Patty

  Ⓒ Miguel

  Ⓓ Chad

5. Justin bought 6 ribbons for an art project. Each ribbon is $\frac{1}{4}$ yard long. How many yards of ribbon did Justin buy? **(Lesson 8.1)**

  Ⓐ $\frac{2}{3}$ yard

  Ⓑ $1\frac{1}{4}$ yards

  Ⓒ $1\frac{1}{2}$ yards

  Ⓓ $1\frac{3}{4}$ yards

6. Kyle and Andrea were asked to make a list of prime numbers.

    Kyle: 1, 3, 7, 19, 23

    Andrea: 2, 3, 5, 7, 11

Whose list is correct? **(Lesson 5.5)**

  Ⓐ Only Kyle's list

  Ⓑ Only Andrea's list

  Ⓒ Both lists are correct.

  Ⓓ Neither list is correct.

Name _____

Chapter 10 Extra Practice

Lesson 10.1

Draw and label an example of the figure.

1. acute ∠MNP

2. $\overline{QR}$

3. $\vec{TS}$

Lesson 10.2

Classify each triangle. Write *acute*, *right*, or *obtuse*.

1.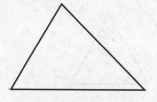
_____

2.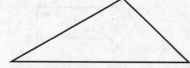
_____

3.
_____

Lesson 10.3

Use the streeet map for 1–2.

1. Name two streets that appear to be parallel.

_____

2. Name two streets that appear to be perpendicular.

_____

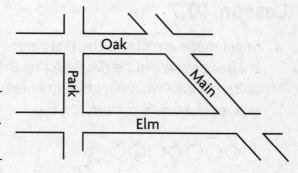

Lesson 10.4

Classify each figure as many ways as possible. Write
quadrilateral, trapezoid, parallelogram, rhombus, rectangle, or *square.*

1.

_____

_____

2.

_____

_____

Lesson 10.5

Tell if the dashed line appears to be a line of symmetry.
Write *yes* or *no*.

1.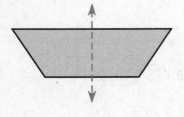

   _____

2.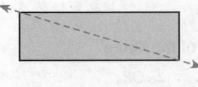

   _____

3.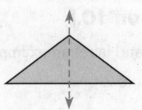

   _____

Lesson 10.6

Does the design have line symmetry? Write *yes* or *no*.
If your answer is *yes*, draw all lines of symmetry.

1.

   _____

2.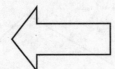

   _____

3.

   _____

Lesson 10.7

1. Sonia made a pattern. The first nine shapes are shown below. Describe the pattern. Draw what might be the next three shapes in Sonia's pattern.

   ○◇○○◇○○◇○ _____

   _____

   _____

   _____

2. Leo makes a pattern with triangles. Draw what might be the next figure in the pattern. How can you describe the pattern?

   ▽△▽△▽△▽ _____

   _____

   _____

   _____

School-Home Letter

© Houghton Mifflin Harcourt Publishing Company

Vocabulary

clockwise The direction the clock hands move

counterclockwise The direction opposite from the way clock hands move

degree (°) A unit for measuring angles

protractor A tool for measuring the size of an angle

Dear Family,

Throughout the next few weeks, our math class will be learning about angles and angle measures. We will also learn to use a protractor to measure and draw angles.

You can expect to see homework in which students find and compute with angle measures.

Here is a sample of how your child will be taught how to relate degrees to fractional parts of a circle.

🔒 MODEL Find Angle Measures

Find the measure of a right angle.

STEP 1

A right angle turns $\frac{1}{4}$ through a circle. Write $\frac{1}{4}$ as an equivalent fraction with 360 in the denominator: $\frac{1}{4} = \frac{90}{360}$

STEP 2

A $\frac{1}{360}$ turn measures 1°. So, a $\frac{90}{360}$ turn measures 90°.

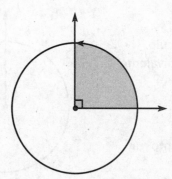

Tips

Classifying Angles

An *acute* angle measures *less than* 90°. An *obtuse* angle measures *more than* 90° and *less than* 180°. A *straight* angle measures 180°.

Activity

Help your child measure angles in pictures of buildings and bridges and decide whether certain angle measures are more common. Then have your child draw his or her own building or bridge design and label each angle measure.

Carta
para la casa

Vocabulario

en el sentido de las manecillas del reloj La dirección en que se mueven las manecillas del reloj

en sentido contrario a las manecillas del reloj La dirección opuesta a cómo se mueven las manecillas del reloj

grado (º) Una unidad para medir los ángulos

transportador Una herramienta para medir el tamaño de un ángulo

Querida familia,

Durante las próximas semanas, en la clase de matemáticas aprenderemos sobre ángulos y medidas de los ángulos. También aprenderemos a usar un transportador y a medir y trazar ángulos.

Llevaré a casa tareas en las que tenga que hallar y hacer cálculos con medidas de ángulos.

Este es un ejemplo de cómo vamos a relacionar los grados con las partes fraccionarias de un círculo.

🔒 MODELO Hallar medidas de ángulos

Halla la medida de un ángulo recto.

PASO 1

Un ángulo recto gira $\frac{1}{4}$ de un círculo. Escribe $\frac{1}{4}$ como una fracción equivalente con 360 en el denominador: $\frac{1}{4} = \frac{90}{360}$

PASO 2

Un giro de $\frac{1}{360}$ mide 1º. Por lo tanto, un giro de $\frac{90}{360}$ mide 90º.

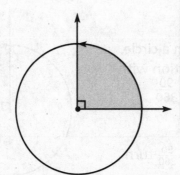

Pistas

Clasificar ángulos

Un ángulo *agudo* mide *menos de 90º*. Un ángulo *obtuso* mide *más de 90º* y *menos de 180º*. Un ángulo *llano* mide 180º.

Actividad

Ayude a su hijo o hija a medir ángulos en dibujos de edificios y puentes y decidan si ciertas medidas de ángulos son más comunes. Luego pídale que dibuje su propio diseño de edificio o puente y que le ponga nombre a cada medida de ángulo.

Name _____

Angles and Fractional Parts of a Circle

Tell what fraction of the circle the shaded angle represents.

1.

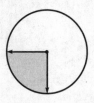

$\dfrac{1}{4}$

_____

2.

_____

3.

_____

Tell whether the angle on the circle shows a $\dfrac{1}{4}, \dfrac{1}{2}, \dfrac{3}{4},$ or 1 full turn clockwise or counterclockwise.

4.

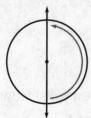

_____

_____

5.

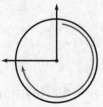

_____

_____

6.

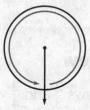

_____

_____

Problem Solving REAL WORLD

7. Shelley exercised for 15 minutes. Describe the turn the minute hand made.

Start

End

_____

_____

8. Mark took 30 minutes to finish lunch. Describe the turn the minute hand made.

Start

End

_____

_____

Lesson Check

1. What fraction of the circle does the shaded angle represent?

Ⓐ $\frac{1}{1}$ or 1   Ⓒ $\frac{1}{2}$

Ⓑ $\frac{3}{4}$ Ⓓ $\frac{1}{4}$

2. Which describes the turn shown below?

Ⓐ $\frac{1}{4}$ turn clockwise

Ⓑ $\frac{1}{2}$ turn clockwise

Ⓒ $\frac{1}{4}$ turn counterclockwise

Ⓓ $\frac{1}{2}$ turn counterclockwise

Spiral Review

3. Which shows $\frac{2}{3}$ and $\frac{3}{4}$ written as a pair of fractions with a common denominator? (Lesson 6.4)

Ⓐ $\frac{2}{3}$ and $\frac{4}{3}$

Ⓑ $\frac{6}{9}$ and $\frac{6}{8}$

Ⓒ $\frac{2}{12}$ and $\frac{3}{12}$

Ⓓ $\frac{8}{12}$ and $\frac{9}{12}$

4. Raymond bought $\frac{3}{4}$ of a dozen rolls. How many rolls did he buy? (Lesson 8.4)

Ⓐ 3

Ⓑ 6

Ⓒ 7

Ⓓ 9

5. Which of the following lists all the factors of 18? (Lesson 5.1)

Ⓐ 1, 2, 4, 9, 18

Ⓑ 1, 2, 3, 6, 9, 18

Ⓒ 2, 3, 6, 9

Ⓓ 1, 3, 5, 9, 18

6. Jonathan rode 1.05 miles on Friday, 1.5 miles on Saturday, 1.25 miles on Monday, and 1.1 miles on Tuesday. On which day did he ride the shortest distance? (Lesson 9.7)

Ⓐ Monday   Ⓒ Friday

Ⓑ Tuesday   Ⓓ Saturday

Degrees

Tell the measure of the angle in degrees.

1.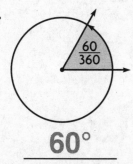

$\dfrac{60}{360}$

60°
_____

2.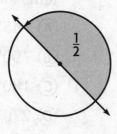

$\dfrac{1}{2}$

_____

3.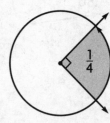

$\dfrac{1}{4}$

_____

Classify the angle. Write *acute*, *obtuse*, *right*, or *straight*.

4.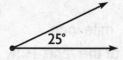

25°

_____

5.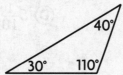

110°

_____

6.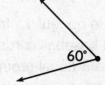

60°

_____

Classify the triangle. Write *acute*, *obtuse*, or *right*.

7.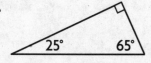

25°   65°

_____

8.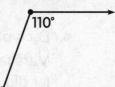

40°
30° 110°

_____

9.

60°   70°
50°

_____

Problem Solving 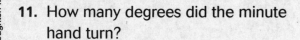 REAL WORLD

Ann started reading at 4:00 P.M. and finished at 4:20 P.M.

10. Through what fraction of a circle did the minute hand turn?

_____

11. How many degrees did the minute hand turn?

_____

Start   **End**

© Houghton Mifflin Harcourt Publishing Company

Lesson Check

1. What kind of angle is shown?

180°

- (A) acute
- (B) obtuse
- (C) right
- (D) straight

2. How many degrees are in an angle that turns through $\frac{1}{4}$ of a circle?

- (A) 45°
- (B) 90°
- (C) 180°
- (D) 270°

Spiral Review

3. Mae bought 15 football cards and 18 baseball cards. She separated them into 3 equal groups. How many sports cards are in each group? **(Lesson 4.12)**

- (A) 5
- (B) 6
- (C) 11
- (D) 12

4. Each part of a race is $\frac{1}{10}$ mile long. Marsha finished 5 parts of the race. How far did Marsha race? **(Lesson 8.1)**

- (A) $\frac{1}{10}$ mile
- (B) $\frac{5}{12}$ mile
- (C) $\frac{1}{2}$ mile
- (D) $5\frac{1}{10}$ miles

5. Jeff said his city got $\frac{11}{3}$ inches of snow. Which shows this fraction written as a mixed number? **(Lesson 7.6)**

- (A) $3\frac{2}{3}$
- (B) $3\frac{1}{3}$
- (C) $2\frac{2}{3}$
- (D) $1\frac{2}{3}$

6. Amy ran $\frac{3}{4}$ mile. Which decimal shows how many miles she ran? **(Lesson 9.3)**

- (A) 0.25 mile
- (B) 0.34 mile
- (C) 0.5 mile
- (D) 0.75 mile

Measure and Draw Angles

Use a protractor to find the angle measure.

1.
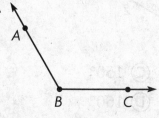

m∠ABC = __**120°**__

2.
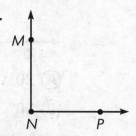

m∠MNP = _____

3.
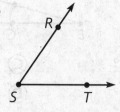

m∠RST = _____

Use a protractor to draw the angle.

4. 40°

5. 170°

Draw an example of each. Label the angle with its measure.

6. a right angle

7. an acute angle

Problem Solving REAL WORLD

The drawing shows the angles a stair tread makes with a support board along a wall. Use your protractor to measure the angles.

8. What is the measure of ∠A? _____

9. What is the measure of ∠B? _____

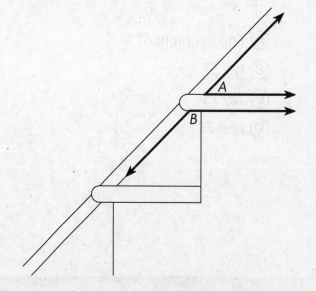

Lesson Check

1. What is the measure of ∠ABC?

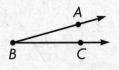

- (A) 15°
- (B) 25°
- (C) 155°
- (D) 165°

2. What is the measure of ∠XYZ?

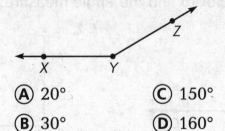

- (A) 20°
- (B) 30°
- (C) 150°
- (D) 160°

Spiral Review

3. Derrick earned $1,472 during the 4 weeks he had his summer job. If he earned the same amount each week, how much did he earn each week?

(Lesson 4.10)

- (A) $360
- (B) $368
- (C) $3,680
- (D) $5,888

4. Arthur baked $1\frac{7}{12}$ dozen muffins. Nina baked $1\frac{1}{12}$ dozen muffins. How many dozen muffins did they bake in all?

(Lesson 7.7)

- (A) $3\frac{2}{3}$
- (B) $2\frac{2}{3}$
- (C) $2\frac{1}{2}$
- (D) $\frac{6}{12}$

5. Trisha drew the figure below. What figure did she draw? **(Lesson 10.1)**

- (A) line segment ST
- (B) ray ST
- (C) ray TS
- (D) line TS

6. Which describes the turn shown by the angle? **(Lesson 11.1)**

- (A) 1 full turn clockwise
- (B) $\frac{3}{4}$ turn clockwise
- (C) $\frac{1}{2}$ turn clockwise
- (D) $\frac{1}{4}$ turn clockwise

Join and Separate Angles

Add to find the measure of the angle. Write an
equation to record your work.

1.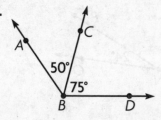

2.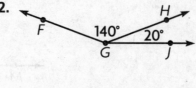

3.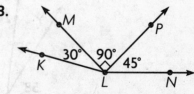

$$50° + 75° = 125°$$

m∠ABD = ____125°____   m∠FGJ = _____   m∠KLN = _____

Use a protractor to find the measure of each
angle in the circle.

4. m∠ABC = _____   5. m∠DBE = _____

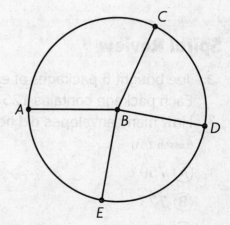

6. m∠CBD = _____   7. m∠EBA = _____

8. Write the sum of the angle measures as an equation.

_____

Problem Solving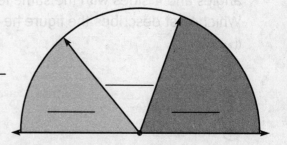

9. Ned made the design at the right. Use a protractor.
   Find and write the measure of each of the 3 angles.

   _____

10. Write an equation to find the measure of the
    total angle.

   _____

Lesson Check

1. What is the measure of ∠WXZ?

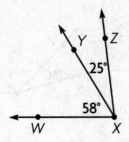

Ⓐ 32°

Ⓑ 83°

Ⓒ 88°

Ⓓ 97°

2. Which equation can you use to find the m∠MNQ?

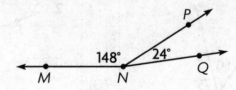

Ⓐ 148° − 24° = ▨

Ⓑ 148° × 24° = ▨

Ⓒ 148° ÷ 24° = ▨

Ⓓ 148° + 24° = ▨

Spiral Review

3. Joe bought 6 packages of envelopes. Each package contains 125 envelopes. How many envelopes did he buy?

(Lesson 2.11)

Ⓐ 750

Ⓑ 723

Ⓒ 720

Ⓓ 650

4. The Lake Trail is $\frac{3}{10}$ mile long and the Rock Trail is $\frac{5}{10}$ long. Bill hiked each trail once. How many miles did he hike in all?

(Lesson 7.5)

Ⓐ $\frac{1}{5}$ mile

Ⓑ $\frac{4}{10}$ mile

Ⓒ $\frac{1}{2}$ mile

Ⓓ $\frac{8}{10}$ mile

5. Ron drew a quadrilateral with 4 right angles and 4 sides with the same length. Which best describes the figure he drew?

(Lesson 10.4)

Ⓐ square

Ⓑ rhombus

Ⓒ trapezoid

Ⓓ parallelogram

6. How many degrees are in an angle that turns through $\frac{3}{4}$ of a circle? (Lesson 11.2)

Ⓐ 45°

Ⓑ 90°

Ⓒ 180°

Ⓓ 270°

Problem Solving • Unknown
Angle Measures

Solve each problem. Draw a diagram to help.

1. Wayne is building a birdhouse. He is cutting a
   board as shown. What is the angle measure of
   the piece left over?

   Draw a bar model to represent the problem.

   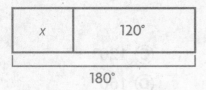

   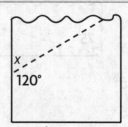

   $$x + 120° = 180°$$
   $$x = 180° - 120°$$
   $$x = 60°$$

   60°
   _____

2. An artist is cutting a piece of metal as shown.
   What is the angle measure of the piece left over?

   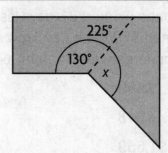

   _____

3. Joan has a piece of material for making a
   costume. She needs to cut it as shown. What is
   the angle measure of the piece left over?

   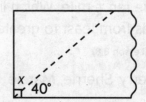

   _____

Lesson Check

1. Angelo cuts a triangle from a sheet of paper as shown. What is the measure of ∠x in the triangle?

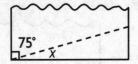

- (A) 15°
- (B) 25°
- (C) 75°
- (D) 105°

2. Cindy cuts a piece of wood as shown. What is the angle measure of the piece left over?

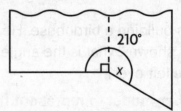

- (A) 30°
- (C) 120°
- (B) 90°
- (D) 150°

Spiral Review

3. Tyronne worked 21 days last month. He earned $79 each day. How much did Tyronne earn last month? (Lesson 3.7)

- (A) $869
- (B) $948
- (C) $1,659
- (D) $2,169

4. Meg inline skated for $\frac{7}{10}$ mile. Which shows this distance written as a decimal? (Lesson 9.1)

- (A) 0.07 mile
- (B) 0.1 mile
- (C) 0.7 mile
- (D) 7.1 miles

5. Kerry ran $\frac{3}{4}$ mile. Sherrie ran $\frac{1}{2}$ mile. Marcie ran $\frac{2}{3}$ mile. Which list orders the friends from least to greatest distance run? (Lesson 6.8)

- (A) Kerry, Sherrie, Marcie
- (B) Kerry, Marcie, Sherrie
- (C) Sherrie, Kerry, Marcie
- (D) Sherrie, Marcie, Kerry

6. What is the measure of ∠ABC? (Lesson 11.4)

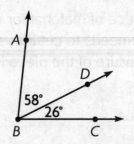

- (A) 32°
- (C) 88°
- (B) 84°
- (D) 94°

Chapter 11 Extra Practice

Lesson 11.1

Tell whether the angle on the circle shows $\frac{1}{4}$, $\frac{1}{2}$, $\frac{3}{4}$, or
1 full turn clockwise or counterclockwise.

1.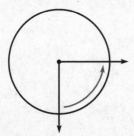

_____

_____

2.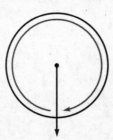

_____

_____

3.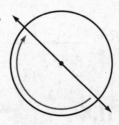

_____

_____

Lesson 11.2

Tell the measure of the angle in degrees.

1.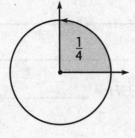

$\frac{1}{4}$

_____

2.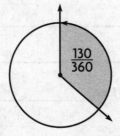

$\frac{130}{360}$

_____

3.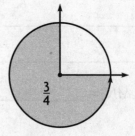

$\frac{3}{4}$

_____

Classify the triangle. Write *acute*, *obtuse*, or *right*.

4.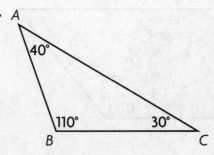

_____

5.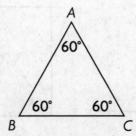

_____

6.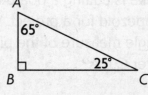

_____

Lesson 11.3

1. Use a protractor to find the angle measure.

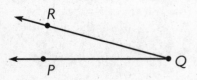

m∠PQR = _____

2. Use a protractor to draw an angle with the measure 72°.

Lesson 11.4

Add to find the measure of the angle. Write an equation to record your work.

1.

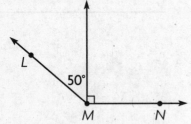

_____

_____

m∠LMN = _____

2.

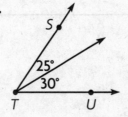

_____

_____

m∠STU = _____

3.

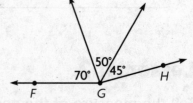

_____

_____

m∠FGH = _____

Lesson 11.5

Use the diagram for 1–2.

1. Luke is cutting a board to make a trapezoid for a project. What is the angle measure of the piece left over after Cut A?

_____

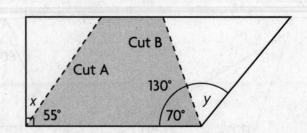

2. What is the angle measure of the piece left over after Cut B?

_____

School-Home Letter

© Houghton Mifflin Harcourt Publishing Company

Vocabulary

decimeter (dm) A metric unit for measuring length or distance

fluid ounce (fl oz) A customary unit for measuring liquid volume

line plot A graph that shows the frequency of data along a number line

second A small unit of time

Dear Family,

During the next few weeks, our math class will be learning about customary and metric units of length, weight/mass, and liquid volume. We will also find elapsed time and learn to compute with mixed measures.

You can expect to see homework on how to use measurement benchmarks and how to compare units.

Here is a sample of how your child will be taught to compare sizes of metric units of length.

🔒 MODEL Compare the Relative Size of Centimeters and Millimeters

Look at a centimeter ruler.

centimeters

Each labeled mark on the ruler is 1 centimeter.
The small marks between centimeters are millimeters.

1 centimeter = 10 millimeters

1 centimeter is 10 times as long as 1 millimeter.

1 millimeter is $\frac{1}{10}$ or 0.1 of a centimeter.

Tips

Estimating Measures

Use benchmarks to help you estimate measures. For example, the width of your finger is about 1 centimeter.

Activity

Have your child commit basic customary and metric units of measure to memory. Work together to make flash cards with measurement units, and have your child practice relating and comparing units. Use daily activities, such as meals and cooking, as opportunities for practice. For example, "If you start with 1 quart of juice and drink 3 cups, how many cups of juice are left?"

Carta para la casa

Vocabulario

decímetro (dm) Una unidad métrica que se usa para medir longitud o distancia

onza fluida (fl oz) Una unidad usual para medir el volumen líquido

diagrama de puntos Una gráfica que muestra la frecuencia de los datos a lo largo de una recta numérica

segundo Una unidad pequeña de tiempo

Querida familia,

Durante las próxima semanas, en la clase de matemáticas aprenderemos las unidades usuales y métricas de longitud, peso/masa y volumen líquido. También aprenderemos a hallar el tiempo transcurrido y a calcular con medidas mixtas.

Llevaré a la casa tareas con actividades para aprender a usar puntos de referencia para medir y a comparar unidades.

Este es un ejemplo de la manera como aprenderemos a comparar los tamaños de las unidades métricas de longitud.

🔑 MODELO Comparar el tamaño relativo de centímetros y milímetros

Observa la regla dividida en centímetros.

```
|1  2  3  4  5  6  7  8  9  10  11
centimeters
```

Pistas

Estimar medidas

Usa puntos de referencia para estimar medidas. Por ejemplo, tu dedo mide alrededor de 1 centímetro de ancho.

Cada marca señalada en la regla es de 1 centímetro. Las marcas pequeñas entre los centímetros son milímetros. 1 centímetro = 10 milímetros

1 centímetro mide 10 veces más que 1 milímetro.

1 milímetro mide $\frac{1}{10}$ o 0.1 de un centímetro.

Actividad

Pida a su hijo o hija que memorice las unidades básicas usuales y métricas de medida. Trabajen juntos para hacer tarjetas nemotécnicas con las unidades de medida, y pídale que relacione y compare unidades. Aproveche las actividades cotidianas, como las comidas o la cocina, para practicar. Por ejemplo, "Si comienzas con 1 cuarto de jugo y te bebes 3 tazas, ¿cuántas tazas de jugo quedan?"

Name _____

Measurement Benchmarks

Use benchmarks to choose the customary unit you would use to measure each.

1. height of a computer

   foot
   _____

2. weight of a table

   _____

3. length of a semi-truck

   _____

4. the amount of liquid a bathtub holds

   _____

Customary Units	
ounce	yard
pound	mile
inch	gallon
foot	cup

Use benchmarks to choose the metric unit you would use to measure each.

5. mass of a grasshopper

   _____

6. the amount of liquid a water bottle holds

   _____

7. length of a soccer field

   _____

8. length of a pencil

   _____

Metric Units	
milliliter	centimeter
liter	meter
gram	kilometer
kilogram	

Circle the better estimate.

9. mass of a chicken egg

   50 grams   50 kilograms

10. length of a car

   12 miles   12 feet

11. amount of liquid a drinking glass holds

   8 ounces   8 quarts

Complete the sentence. Write *more* or *less*.

12. A camera has a length of _____ than one centimeter.

13. A bowling ball weighs _____ than one pound.

Problem Solving REAL WORLD

14. What is the better estimate for the mass of a textbook, 1 gram or 1 kilogram?

   _____

15. What is the better estimate for the height of a desk, 1 meter or 1 kilometer?

   _____

Lesson Check

1. Which is the best estimate for the weight of a stapler?

   Ⓐ 4 ounces

   Ⓑ 4 pounds

   Ⓒ 4 inches

   Ⓓ 4 feet

2. Which is the best estimate for the length of a car?

   Ⓐ 4 kilometers

   Ⓑ 4 tons

   Ⓒ 4 kilograms

   Ⓓ 4 meters

Spiral Review

3. Bart practices his trumpet $1\frac{1}{4}$ hours each day. How many hours will he practice in 6 days? **(Lesson 8.4)**

   Ⓐ $8\frac{2}{4}$ hours

   Ⓑ $7\frac{2}{4}$ hours

   Ⓒ 7 hours

   Ⓓ $6\frac{2}{4}$ hours

4. Millie collected 100 stamps from different countries. Thirty-two of the stamps are from countries in Africa. What is $\frac{32}{100}$ written as a decimal? **(Lesson 9.2)**

   Ⓐ 32

   Ⓑ 3.2

   Ⓒ 0.32

   Ⓓ 0.032

5. Diedre drew a quadrilateral with 4 right angles and 4 sides of the same length. What kind of polygon did Diedre draw?

   (Lesson 10.4)

   Ⓐ square

   Ⓑ trapezoid

   Ⓒ hexagon

   Ⓓ pentagon

6. How many degrees are in an angle that turns through $\frac{1}{2}$ of a circle? **(Lesson 11.2)**

   Ⓐ 60°

   Ⓑ 90°

   Ⓒ 120°

   Ⓓ 180°

Name _____

Customary Units of Length

Complete.

1. 3 feet = __**36**__ inches   Think: 1 foot = 12 inches,
   so 3 feet = 3 × 12 inches, or 36 inches

2. 2 yards = _____ feet

3. 8 feet = _____ inches

4. 7 yards = _____ feet

5. 4 feet = _____ inches

6. 15 yards = _____ feet

7. 10 feet = _____ inches

Compare using <, >, or =.

8. 3 yards ◯ 10 feet

9. 5 feet ◯ 60 inches

10. 8 yards ◯ 20 feet

11. 3 feet ◯ 10 inches

12. 3 yards ◯ 21 feet

13. 6 feet ◯ 72 inches

Problem Solving REAL WORLD

14. Carla has two lengths of ribbon. One ribbon is 2 feet long. The other ribbon is 30 inches long. Which length of ribbon is longer? **Explain.**

   _____

   _____

   _____

15. A football player gained 2 yards on one play. On the next play, he gained 5 feet. Was his gain greater on the first play or the second play? **Explain.**

   _____

   _____

   _____

Lesson Check

1. Marta has 14 feet of wire to use to make necklaces. She needs to know the length in inches so she can determine how many necklaces to make. How many inches of wire does Marta have?

   (A) 42 inches   (C) 168 inches

   (B) 84 inches   (D) 504 inches

2. Jarod bought 8 yards of ribbon. He needs 200 inches to use to make curtains. How many inches of ribbon does he have?

   (A) 8 inches   (C) 96 inches

   (B) 80 inches   (D) 288 inches

Spiral Review

3. Which describes the turn shown below? **(Lesson 11.1)**

   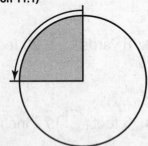

   (A) $\frac{1}{4}$ turn counterclockwise

   (B) $\frac{1}{4}$ turn clockwise

   (C) $\frac{1}{2}$ turn clockwise

   (D) $\frac{3}{4}$ turn counterclockwise

4. Which decimal represents the shaded part of the model below? **(Lesson 9.1)**

   (A) 0.03

   (B) 0.3

   (C) 0.33

   (D) 0.7

5. Three sisters shared $3.60 equally. How much did each sister get? **(Lesson 9.5)**

   (A) $1.00

   (B) $1.20

   (C) $1.80

   (D) $10.80

6. Which is the best estimate for the width of your index finger? **(Lesson 12.1)**

   (A) 1 millimeter

   (B) 1 gram

   (C) 1 centimeter

   (D) 1 liter

Name _____

Customary Units of Weight

Complete.

1. 5 pounds = _____**80**_____ ounces

   Think: 1 pound = 16 ounces, so
   5 pounds = 5 × 16 ounces, or 80 ounces

2. 7 tons = _____ pounds

3. 2 pounds = _____ ounces

4. 3 tons = _____ pounds

5. 10 pounds = _____ ounces

6. 5 tons = _____ pounds

7. 7 pounds = _____ ounces

Compare using <, >, or =.

8. 8 pounds $\bigcirc$ 80 ounces

9. 1 ton $\bigcirc$ 100 pounds

10. 3 pounds $\bigcirc$ 50 ounces

11. 5 tons $\bigcirc$ 1,000 pounds

12. 16 pounds $\bigcirc$ 256 ounces

13. 8 tons $\bigcirc$ 16,000 pounds

Problem Solving REAL WORLD

14. A company that makes steel girders can produce 6 tons of girders in one day. How many pounds is this?

15. Larry's baby sister weighed 6 pounds at birth. How many ounces did the baby weigh?

_____

_____

Lesson Check

1. Ann bought 2 pounds of cheese to make lasagna. The recipe gives the amount of cheese needed in ounces. How many ounces of cheese did she buy?

   Ⓐ 20 ounces

   Ⓑ 32 ounces

   Ⓒ 40 ounces

   Ⓓ 64 ounces

2. A school bus weighs 7 tons. The weight limit for a bridge is given in pounds. What is this weight of the bus in pounds?

   Ⓐ 700 pounds

   Ⓑ 1,400 pounds

   Ⓒ 7,000 pounds

   Ⓓ 14,000 pounds

Spiral Review

3. What is the measure of ∠EHG?
   (Lesson 11.3)

   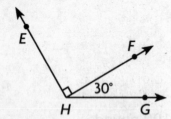

   Ⓐ 60°   Ⓒ 120°

   Ⓑ 100°   Ⓓ 130°

4. How many lines of symmetry does the square below have? (Lesson 10.6)

   Ⓐ 0   Ⓒ 4

   Ⓑ 2   Ⓓ 6

5. To make dough, Reba needs $2\frac{1}{2}$ cups of flour. How much flour does she need to make 5 batches of dough? (Lesson 8.4)

   Ⓐ $14\frac{1}{2}$ cups

   Ⓑ $12\frac{1}{2}$ cups

   Ⓒ $11\frac{1}{2}$ cups

   Ⓓ $10\frac{1}{2}$ cups

6. Judi's father is 6 feet tall. The minimum height to ride a rollercoaster is given in inches. How many inches tall is Judi's father? (Lesson 12.2)

   Ⓐ 60 inches

   Ⓑ 66 inches

   Ⓒ 72 inches

   Ⓓ 216 inches

Name _____

Customary Units of Liquid Volume

Complete.

1. 6 gallons = __24__ quarts

Think: 1 gallon = 4 quarts,
so 6 gallons = 6 × 4 quarts, or 24 quarts

2. 12 quarts = _____ pints

3. 6 cups = _____ fluid ounces

4. 9 pints = _____ cups

5. 10 quarts = _____ cups

6. 5 gallons = _____ pints

7. 3 gallons = _____ cups

Compare using <, >, or =.

8. 6 pints ◯ 60 fluid ounces

9. 3 gallons ◯ 30 quarts

10. 5 quarts ◯ 20 cups

11. 6 cups ◯ 12 pints

12. 8 quarts ◯ 16 pints

13. 6 gallons ◯ 96 pints

Problem Solving REAL WORLD

14. A chef makes $1\frac{1}{2}$ gallons of soup in a large pot. How many 1-cup servings can the chef get from this large pot of soup?

15. Kendra's water bottle contains 2 quarts of water. She wants to add drink mix to it, but the directions for the drink mix give the amount of water in fluid ounces. How many fluid ounces are in her bottle?

© Houghton Mifflin Harcourt Publishing Company

Lesson Check

1. Joshua drinks 8 cups of water a day. The recommended daily amount is given in fluid ounces. How many fluid ounces of water does he drink each day?

(A) 16 fluid ounces

(B) 32 fluid ounces

(C) 64 fluid ounces

(D) 128 fluid ounces

2. A cafeteria used 5 gallons of milk in preparing lunch. How many 1-quart containers of milk did the cafeteria use?

(A) 10

(B) 20

(C) 40

(D) 80

Spiral Review

3. Roy uses $\frac{1}{4}$ cup of batter for each muffin. Which list shows the amounts of batter he will use depending on the number of muffins he makes? (Lesson 8.1)

(A) $\frac{1}{4}, \frac{1}{5}, \frac{1}{6}, \frac{1}{7}, \frac{1}{8}$

(B) $\frac{1}{4}, \frac{2}{4}, \frac{3}{4}, \frac{4}{4}, \frac{5}{4}$

(C) $\frac{1}{4}, \frac{2}{8}, \frac{3}{12}, \frac{4}{16}, \frac{5}{20}$

(D) $\frac{1}{4}, \frac{2}{8}, \frac{4}{16}, \frac{6}{24}, \frac{8}{32}$

4. Beth has $\frac{7}{100}$ of a dollar. Which shows the amount of money Beth has? (Lesson 9.4)

(A) $7.00

(B) $0.70

(C) $0.07

(D) $0.007

5. Name the figure that Enrico drew below. (Lesson 10.1)

(A) a ray

(B) a line

(C) a line segment

(D) an octagon

6. A hippopotamus weighs 4 tons. Feeding instructions are given for weights in pounds. How many pounds does the hippopotamus weigh? (Lesson 12.3)

(A) 4,000 pounds

(B) 6,000 pounds

(C) 8,000 pounds

(D) 12,000 pounds

Name _____

Line Plots

1. Some students compared the time they spend riding the school bus. Complete the tally table and line plot to show the data.

Time Spent on School Bus	
Time (in hours)	Tally
$\frac{1}{6}$	| |
$\frac{2}{6}$	
$\frac{3}{6}$	
$\frac{4}{6}$	

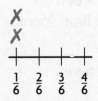

Time Spent on School Bus
(in hours)

$$\frac{1}{6}, \frac{3}{6}, \frac{4}{6}, \frac{2}{6}, \frac{3}{6}, \frac{1}{6}, \frac{3}{6}, \frac{3}{6}$$

X
X
$\frac{1}{6}$ $\frac{2}{6}$ $\frac{3}{6}$ $\frac{4}{6}$

Time Spent on School Bus (in hours)

Use your line plot for 2 and 3.

2. How many students compared times? _____

3. What is the difference between the longest time and shortest

   time students spent riding the bus? _____

Problem Solving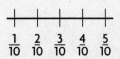

For 4–5, make a tally table on a separate sheet of paper.
Make a line plot in the space below the problem.

4.
Milk Drunk at Lunch
(in quarts)

$$\frac{1}{8}, \frac{2}{8}, \frac{2}{8}, \frac{4}{8}, \frac{1}{8}, \frac{3}{8}, \frac{4}{8}, \frac{2}{8}, \frac{3}{8}, \frac{2}{8}$$

5.
Distance Between Stops for a Rural
Mail Carrier (in miles)

$$\frac{3}{10}, \frac{4}{10}, \frac{5}{10}, \frac{1}{10}, \frac{5}{10}, \frac{4}{10}, \frac{4}{10}, \frac{3}{10}$$

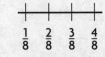

$\frac{1}{8}$ $\frac{2}{8}$ $\frac{3}{8}$ $\frac{4}{8}$

Milk Drunk at Lunch
(in quarts)

$\frac{1}{10}$ $\frac{2}{10}$ $\frac{3}{10}$ $\frac{4}{10}$ $\frac{5}{10}$

Distance Between Stops for
a Rural Mail Carrier (in miles)

Lesson Check

Use the line plot for 1 and 2.

1. How many students were reading during study time?

   Ⓐ 5   Ⓒ 7

   Ⓑ 6   Ⓓ 8

2. What is the difference between the longest time and shortest time spent reading?

   Ⓐ $\frac{4}{8}$ hour   Ⓒ $\frac{2}{8}$ hour

   Ⓑ $\frac{3}{8}$ hour   Ⓓ $\frac{1}{8}$ hour

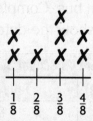

Time Spent Reading During Study Time (in hours)

Spiral Review

3. Bridget is allowed to play on-line games for $\frac{75}{100}$ of an hour each day. Which shows that fraction as a decimal? **(Lesson 9.3)**

   Ⓐ 75.0

   Ⓑ 7.50

   Ⓒ 0.75

   Ⓓ 0.075

4. Bobby's collection of sports cards has $\frac{3}{10}$ baseball cards and $\frac{39}{100}$ football cards. The rest are soccer cards. What fraction of Bobby's sports cards are baseball or football cards? **(Lesson 9.6)**

   Ⓐ $\frac{9}{100}$ Ⓒ $\frac{52}{100}$

   Ⓑ $\frac{42}{100}$ Ⓓ $\frac{69}{100}$

5. Jeremy gives his horse 12 gallons of water each day. How many 1-quart pails of water is that? **(Lesson 12.4)**

   Ⓐ 24   Ⓒ 72

   Ⓑ 48   Ⓓ 96

6. An iguana at a pet store is 5 feet long. Measurements for iguana cages are given in inches. How many inches long is the iguana? **(Lesson 12.2)**

   Ⓐ 45 inches   Ⓒ 60 inches

   Ⓑ 50 inches   Ⓓ 72 inches

Name _____

Metric Units of Length

Complete.

1. 4 meters = ___**400**___ centimeters

Think: 1 meter = 100 centimeters,
so 4 meters = 4 × 100 centimeters,
or 400 centimeters

2. 8 centimeters = _____ millimeters

3. 5 meters = _____ decimeters

4. 9 meters = _____ millimeters

5. 7 meters = _____ centimeters

Compare using <, >, or =.

6. 8 meters ◯ 80 centimeters

7. 3 decimeters ◯ 30 centimeters

8. 4 meters ◯ 450 centimeters

9. 90 centimeters ◯ 9 millimeters

Describe the length in meters. Write your answer as a fraction and as a decimal.

10. 43 centimeters = _____ or

_____ meter

11. 6 decimeters = _____ or

_____ meter

12. 8 centimeters = _____ or

_____ meter

13. 3 decimeters = _____ or

_____ meter

Problem Solving REAL WORLD

14. A flagpole is 4 meters tall. How many centimeters tall is the flagpole?

_____

15. A new building is 25 meters tall. How many decimeters tall is the building?

_____

Lesson Check

1. A pencil is 15 centimeters long. How many millimeters long is that pencil?

  (A)   1.5 millimeters

  (B)   15 millimeters

  (C)   150 millimeters

  (D) 1,500 millimeters

2. John's father is 2 meters tall. How many centimeters tall is John's father?

  (A) 2,000 centimeters

  (B)   200 centimeters

  (C)   20 centimeters

  (D)   2 centimeters

Spiral Review

3. Bruce reads for $\frac{3}{4}$ hour each night. How long will he read in 4 nights? **(Lesson 8.3)**

  (A) $\frac{3}{16}$ hour

  (B) $\frac{7}{4}$ hours

  (C) $\frac{9}{4}$ hours

  (D) $\frac{12}{4}$ hours

4. Mark jogged 0.6 mile. Caroline jogged 0.49 mile. Which inequality correctly compares the distances they jogged?

(Lesson 9.7)

  (A) $0.6 = 0.49$

  (B) $0.6 > 0.49$

  (C) $0.6 < 0.49$

  (D) $0.6 + 0.49 = 1.09$

Use the line plot for 5 and 6.

5. How many lawns were mowed? **(Lesson 12.5)**

  (A) 8   (C) 10

  (B) 9   (D) 11

6. What is the difference between the greatest amount and least amount of gasoline used to mow lawns? **(Lesson 12.5)**

  (A) $\frac{6}{8}$ gallon   (C) $\frac{4}{8}$ gallon

  (B) $\frac{5}{8}$ gallon   (D) $\frac{3}{8}$ gallon

```
                X
        X       X
        X   X   X
    X   X   X   X   X
    +---+---+---+---+
    1   2   3   4   5
    8   8   8   8   8
```

Gasoline Used to Mow Lawns in May (in Gallons)

Name _____

Metric Units of Mass and Liquid Volume

Complete.

1. 5 liters = **5,000** milliliters

Think: 1 liter = 1,000 milliliters,
so 5 liters = 5 × 1,000 milliliters, or 5,000 milliliters

2. 3 kilograms = _____ grams

3. 8 liters = _____ milliliters

4. 7 kilograms = _____ grams

5. 9 liters = _____ milliliters

6. 2 liters = _____ milliliters

7. 6 kilograms = _____ grams

Compare using <, >, or =.

8. 8 kilograms $\bigcirc$ 850 grams

9. 3 liters $\bigcirc$ 3,500 milliliters

10. 1 kilogram $\bigcirc$ 1,000 grams

11. 5 liters $\bigcirc$ 520 milliliters

Problem Solving  REAL WORLD

12. Kenny buys four 1-liter bottles of water. How many milliliters of water does Kenny buy?

_____

13. Mrs. Jones bought three 2-kilogram packages of flour. How many grams of flour did she buy?

_____

14. Colleen bought 8 kilograms of apples and 2.5 kilograms of pears. How many more grams of apples than pears did she buy?

_____

15. Dave uses 500 milliliters of juice for a punch recipe. He mixes it with 2 liters of ginger ale. How many milliliters of punch does he make?

_____

Lesson Check

1. During his hike, Milt drank 1 liter of water and 1 liter of sports drink. How many milliliters of liquid did he drink in all?

   (A) 20 milliliters

   (B) 200 milliliters

   (C) 2,000 milliliters

   (D) 20,000 milliliters

2. Larinda cooked a 4-kilogram roast. The roast left over after the meal weighed 3 kilograms. How many grams of roast were eaten during that meal?

   (A) 7,000 grams

   (B) 1,000 grams

   (C) 700 grams

   (D) 100 grams

Spiral Review

3. Use a protractor to find the angle measure. (Lesson 11.3)

   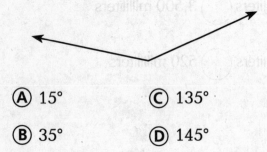

   (A) 15°   (C) 135°

   (B) 35°   (D) 145°

4. Which of the following shows parallel lines? (Lesson 10.3)

   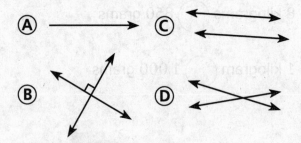

5. Carly bought 3 pounds of birdseed. How many ounces of birdseed did she buy?

   (Lesson 12.3)

   (A) 30 ounces

   (B) 36 ounces

   (C) 42 ounces

   (D) 48 ounces

6. A door is 8 decimeters wide. How wide is the door in centimeters? (Lesson 12.6)

   (A) 8 centimeters

   (B) 80 centimeters

   (C) 800 centimeters

   (D) 8,000 centimeters

Units of Time

Complete.

1. 6 minutes = _____**360**_____ seconds   Think: 1 minute = 60 seconds,
so 6 minutes = 6 × 60 seconds, or 360 seconds

2. 5 weeks = _____ days

3. 3 years = _____ weeks

4. 9 hours = _____ minutes

5. 9 minutes = _____ seconds

6. 5 years = _____ months

7. 7 days = _____ hours

Compare using <, >, or =.

8. 2 years ◯ 14 months

9. 3 hours ◯ 300 minutes

10. 2 days ◯ 48 hours

11. 6 years ◯ 300 weeks

12. 4 hours ◯ 400 minutes

13. 5 minutes ◯ 300 seconds

Problem Solving `REAL WORLD`

14. Jody practiced a piano piece for 500 seconds. Bill practiced a piano piece for 8 minutes. Who practiced longer? **Explain.**

_____

_____

_____

15. Yvette's younger brother just turned 3 years old. Fred's brother is now 30 months old. Whose brother is older? **Explain.**

_____

_____

_____

Lesson Check

1. Glen rode his bike for 2 hours. For how many minutes did Glen ride his bike?

   (A) 60 minutes

   (B) 100 minutes

   (C) 120 minutes

   (D) 150 minutes

2. Tina says that vacation starts in exactly 4 weeks. In how many days does vacation start?

   (A) 28 days

   (B) 35 days

   (C) 42 days

   (D) 48 days

Spiral Review

3. Kayla bought $\frac{9}{4}$ pounds of apples. What is that weight as a mixed number? (Lesson 7.6)

   (A) $1\frac{1}{4}$ pounds

   (B) $1\frac{4}{9}$ pounds

   (C) $2\frac{1}{4}$ pounds

   (D) $2\frac{3}{4}$ pounds

4. Judy, Jeff, and Jim each earned $5.40 raking leaves. How much did they earn in all? (Lesson 9.5)

   (A) $1.60

   (B) $10.80

   (C) $15.20

   (D) $16.20

5. Melinda rode her bike $\frac{54}{100}$ mile to the library. Then she rode $\frac{4}{10}$ mile to the store. How far did Melinda ride her bike in all? (Lesson 9.6)

   (A) 0.14 mile

   (B) 0.58 mile

   (C) 0.94 mile

   (D) 1.04 miles

6. One day, the students drank 60 quarts of milk at lunch. How many pints of milk did the students drink? (Lesson 12.4)

   (A) 30 pints

   (B) 120 pints

   (C) 240 pints

   (D) 480 pints

Name _____

Problem Solving · Elapsed Time

Read each problem and solve.

1. Molly started her piano lesson at 3:45 P.M. The lesson lasted 20 minutes. What time did the piano lesson end?

   Think: What do I need to find? How can I draw a diagram to help?

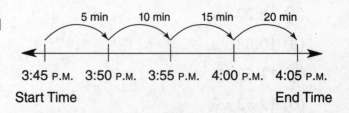

   4:05 P.M.

2. Brendan spent 24 minutes playing a computer game. He stopped playing at 3:55 P.M and went outside to ride his bike. What time did he start playing the computer game?

3. Aimee's karate class lasts 1 hour and 15 minutes and is over at 5:00 P.M. What time does Aimee's karate class start?

4. Mr. Giarmo left for work at 7:15 A.M. Twenty-five minutes later, he arrived at his work. What time did Mr. Giarmo arrive at his work?

5. Ms. Brown's flight left at 9:20 A.M. Her plane landed 1 hour and 23 minutes later. What time did her plane land?

Lesson Check

1. Bobbie went snowboarding with friends at 10:10 A.M. They snowboarded for 1 hour and 43 minutes, and then stopped to eat lunch. What time did they stop for lunch?

   (A) 8:27 A.M.

   (B) 10:53 A.M.

   (C) 11:53 A.M.

   (D) 12:53 A.M.

2. The Cain family drove for 1 hour and 15 minutes and arrived at their camping spot at 3:44 P.M. What time did the Cain family start driving?

   (A) 4:59 P.M.

   (B) 2:44 P.M.

   (C) 2:39 P.M.

   (D) 2:29 P.M.

Spiral Review

3. A praying mantis can grow up to 15 centimeters long. How long is this in millimeters? (Lesson 12.6)

   (A) 15 millimeters

   (B) 150 millimeters

   (C) 1,500 millimeters

   (D) 15,000 millimeters

4. Thom's minestrone soup recipe makes 3 liters of soup. How many milliliters of soup is this? (Lesson 12.7)

   (A) 30 milliliters

   (B) 300 milliliters

   (C) 3,000 milliliters

   (D) 30,000 milliliters

5. Stewart walks $\frac{2}{3}$ mile each day. Which is a multiple of $\frac{2}{3}$? (Lesson 8.2)

   (A) $\frac{4}{3}$

   (B) $\frac{4}{6}$

   (C) $\frac{8}{10}$

   (D) $\frac{2}{12}$

6. Angelica colored in 0.60 of the squares on her grid. Which of the following expresses 0.60 as tenths in fraction form? (Lesson 9.3)

   (A) $\frac{60}{100}$

   (B) $\frac{60}{10}$

   (C) $\frac{6}{100}$

   (D) $\frac{6}{10}$

Name _____

Mixed Measures

Complete.

1. 8 pounds 4 ounces = _____**132**_____ ounces

Think: 8 pounds = 8 × 16 ounces, or 128 ounces.

128 ounces + 4 ounces = 132 ounces

2. 5 weeks 3 days = _____ days

3. 4 minutes 45 seconds = _____ seconds

4. 4 hours 30 minutes = _____ minutes

5. 3 tons 600 pounds = _____ pounds

6. 6 pints 1 cup = _____ cups

7. 7 pounds 12 ounces = _____ ounces

Add or subtract.

8. 9 gal 1 qt
  + 6 gal 1 qt
  _____

9. 12 lb 5 oz
  − 7 lb 10 oz
  _____

10. 8 hr 3 min
  + 4 hr 12 min
  _____

Problem Solving REAL WORLD

11. Michael's basketball team practiced for 2 hours 40 minutes yesterday and 3 hours 15 minutes today. How much longer did the team practice today than yesterday?

12. Rhonda had a piece of ribbon that was 5 feet 3 inches long. She removed a 5-inch piece to use in her art project. What is the length of the piece of ribbon now?

Lesson Check

1. Marsha bought 1 pound 11 ounces of roast beef and 2 pounds 5 ounces of corned beef. How much more corned beef did she buy than roast beef?

Ⓐ 16 ounces

Ⓑ 10 ounces

Ⓒ 7 ounces

Ⓓ 6 ounces

2. Theodore says there are 2 weeks 5 days left in the year. How many days are left in the year?

Ⓐ 14 days

Ⓑ 15 days

Ⓒ 19 days

Ⓓ 25 days

Spiral Review

3. On one grid, 0.5 of the squares are shaded. On another grid, 0.05 of the squares are shaded. Which statement is true? **(Lesson 9.7)**

Ⓐ $0.05 > 0.5$

Ⓑ $0.05 = 0.5$

Ⓒ $0.05 < 0.5$

Ⓓ $0.05 + 0.5 = 1.0$

4. Classify the triangle shown below. **(Lesson 10.2)**

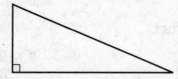

Ⓐ right

Ⓑ acute

Ⓒ equilateral

Ⓓ obtuse

5. Sahil's brother is 3 years old. How many weeks old is his brother? **(Lesson 12.8)**

Ⓐ 30 weeks

Ⓑ 36 weeks

Ⓒ 90 weeks

Ⓓ 156 weeks

6. Sierra's swimming lessons last 1 hour 20 minutes. She finished her lesson at 10:50 A.M. At what time did her lesson start? **(Lesson 12.9)**

Ⓐ 9:30 A.M.

Ⓑ 9:50 A.M.

Ⓒ 10:30 A.M.

Ⓓ 12:10 A.M.

Patterns in Measurement Units

Each table shows a pattern for two customary units
of time or volume. Label the columns of the table.

1.

Gallons	Quarts
1	4
2	8
3	12
4	16
5	20

2.

___	___
1	12
2	24
3	36
4	48
5	60

3.

___	___
1	2
2	4
3	6
4	8
5	10

4.

___	___
1	7
2	14
3	21
4	28
5	35

Problem Solving REAL WORLD

Use the table for 5 and 6.

5. Marguerite made the table to compare two
metric measures of length. Name a pair of
units Marguerite could be comparing.

_____

6. Name another pair of metric units of
length that have the same relationship.

_____

?	?
1	10
2	20
3	30
4	40
5	50

Lesson Check

1. Joanne made a table to relate two units of measure. The number pairs in her table are 1 and 16, 2 and 32, 3 and 48, 4 and 64. Which are the best labels for Joanne's table?

   (A) Cups, Fluid Ounces

   (B) Gallons, Quarts

   (C) Pounds, Ounces

   (D) Yards, Inches

2. Cade made a table to relate two units of time. The number pairs in his table are 1 and 24, 2 and 48, 3 and 72, 4 and 96. Which are the best labels for Cade's table?

   (A) Days, Hours

   (B) Days, Weeks

   (C) Years, Months

   (D) Years, Weeks

Spiral Review

3. Anita has 2 quarters, 1 nickel, and 4 pennies. Write Anita's total amount as a fraction of a dollar. (Lesson 9.4)

   (A) $\frac{39}{100}$

   (B) $\frac{54}{100}$

   (C) $\frac{59}{100}$

   (D) $\frac{84}{100}$

4. The minute hand of a clock moves from 12 to 6. Which describes the turn the minute hand makes? (Lesson 11.1)

   (A) $\frac{1}{4}$ turn

   (B) $\frac{1}{2}$ turn

   (C) $\frac{3}{4}$ turn

   (D) 1 full turn

5. Roderick has a dog that has a mass of 9 kilograms. What is the mass of the dog in grams? (Lesson 12.7)

   (A) 9 grams

   (B) 900 grams

   (C) 9,000 grams

   (D) 90,000 grams

6. Kari mixed 3 gallons 2 quarts of lemon-lime drink with 2 gallons 3 quarts of pink lemonade to make punch. How much more lemon-lime drink did Kari use than pink lemonade? (Lesson 12.10)

   (A) 3 quarts

   (B) 4 quarts

   (C) 1 gallon 1 quart

   (D) 1 gallon 2 quarts

Name _____

Chapter 12 Extra Practice

Lesson 12.1

Use benchmarks to choose the unit you would use to measure each.

1. length of a car

   customary unit: _____

   metric unit: _____

2. liquid volume of a sink

   customary unit: _____

   metric unit: _____

3. weight or mass of a parakeet

   customary unit: _____

   metric unit: _____

4. length of your thumb

   customary unit: _____

   metric unit: _____

Lessons 12.2 – 12.4

Complete.

1. 6 yards = _____ feet

2. 2 feet = _____ inches

3. 3 pounds = _____ ounces

4. 2 tons = _____ pounds

5. 5 gallons = _____ quarts

6. 4 quarts = _____ cups

Lesson 12.5

Use the line plot for 1–2.

1. What is the difference in height between the tallest plant and the shortest plant?

   _____

2. How many plants are in Box A? _____

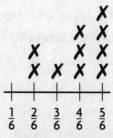

Plant Heights in Box A (in feet)

Lessons 12.6 – 12.8

Complete.

1. 9 centimeters = _____ millimeters

2. 7 meters = _____ decimeters

3. 5 decimeters = _____ centimeters

4. 4 liters = _____ milliliters

5. 3 kilograms = _____ grams

6. 3 weeks = _____ days

7. 6 hours = _____ minutes

8. 2 days = _____ hours

Lesson 12.10

Add or subtract.

1. 3 ft 8 in.	2. 9 lb 6 oz	3. 5 gal 2 qt	4. 7 hr 10 min
+ 1 ft 2 in.	− 4 lb 2 oz	− 1 gal 3 qt	− 3 hr 40 min

Lessons 12.9 and 12.11

1. Rick needs to be at school at 8:15 A.M. It takes him 20 minutes to walk to school. At what time does he need to leave to get to school on time?

2. Sunny's gymnastics class lasts 1 hour 20 minutes. The class starts at 3:50 P.M. At what time does the gymnastics class end?

_____

_____

3. David made a table to relate two customary units. Label the columns of the table.

1	16
2	32
3	48
4	64
5	80

School-Home Letter

Vocabulary

area The number of square units needed to cover a flat surface

base, *b* A polygon's side

formula A set of symbols that expresses a mathematical rule

height, *h* The length of a perpendicular from the base to the top of a two-dimensional figure

perimeter The distance around a figure

square unit A unit of area with dimensions of 1 unit × 1 unit

Dear Family,

During the next few weeks, our math class will be learning about perimeter and area. We will explore the concept that area is a measure of how many square units cover a flat surface. We will also learn the formula for finding the area of a rectangle.

You can expect to see homework that provides practice with finding perimeters and areas of rectangles, and areas of combined rectangles.

Here is a sample of how your child will be taught to use a formula to find the area of a rectangle.

🔑 MODEL Use a Formula to Find Area

This is how we will use a formula to find the area of a rectangle.

STEP 1

Identify the base and the height of the rectangle.

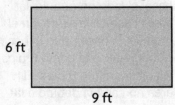

6 ft

9 ft

base = 9 feet

height = 6 feet

STEP 2

Use the formula
$A = b \times h$
to find the area of the rectangle.

$A = 9 \times 6$
$= 54$

The area is 54 square feet.

Tips

Remember that any side of a rectangle could be the base. Depending upon the side labeled as the base, the perpendicular side to that base is the height. In the model, the base could have been identified as 6 feet and the height as 9 feet. Because of the Commutative Property of Multiplication, the area does not change.

Appropriate Units

Remember to use the correct *square* units when expressing the area of a shape. A measure of 54 feet would simply be a measure of length, whereas a measure of 54 *square* feet is a measure of area.

Carta para la casa

Vocabulario

área La cantidad de unidades cuadradas que se necesitan para cubrir una superficie plana

base, b Un lado de un polígono

fórmula Un conjunto de símbolos que expresa una regla matemática

altura, h La longitud de un lado perpendicular de una figura bidimensional desde la base hasta la parte superior

perímetro La distancia alrededor de una figura

unidad cuadrada Una unidad para medir el área que tiene 1 unidad de largo y 1 unidad de ancho

Querida familia,

Durante las próximas semanas, en la clase de matemáticas aprenderemos acerca del perímetro y el área. Exploraremos el concepto del área como medida de superficie que usa unidades cuadradas. También aprenderemos la fórmula para hallar el área de un rectángulo.

Llevaré a la casa tareas para practicar la manera de hallar los perímetros y las áreas de rectángulos y las áreas de combinaciones de rectángulos.

Este es un ejemplo de la manera como aprenderemos a usar una fórmula para hallar el área de un rectángulo.

🔑 MODELO Usar una fórmula para hallar el área

Así es como usaremos la fórmula del área de un rectángulo.

PASO 1

Identifica la base y la altura del rectángulo.

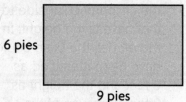

6 pies

9 pies

base = 9 pies
altura = 6 pies

PASO 2

Usa la fórmula
$A = b \times h$
para hallar el área del rectángulo.

$A = 9 \times 6$
$= 54$
El área mide 54 pies cuadrados.

Pistas

Recuerda que cualquiera de los lados de un rectángulo puede ser la base. Según el lado que se determine como base, el lado perpendicular a esa base es la altura. En el modelo, la base pudo haber sido identificada como 6 pies y la altura como 9 pies. El área no cambia debido a la propiedad conmutativa de la multiplicación.

Unidades adecuadas

Recuerda que se debe utilizar la unidad *cuadrada* correcta cuando se expresa el área de una figura. Una medida de 54 pies sería simplemente una medida del largo, en cambio una medida de 54 pies *cuadrados* es una medida del área.

Perimeter

Find the perimeter of the rectangle or square.

1.

3 in.

9 in.

$9 + 3 + 9 + 3 = 24$

24 inches

2.

8 m

8 m

_____ meters

3.

12 ft

10 ft

_____ feet

4.

24 cm

30 cm

_____ centimeters

5.

83 in.

25 in.

_____ inches

6.

60 m

60 m

_____ meters

Problem Solving 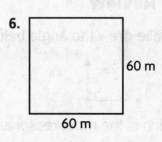 REAL WORLD

7. Troy is making a flag shaped like a square. Each side measures 12 inches. He wants to add ribbon along the edges. He has 36 inches of ribbon. Does he have enough ribbon? **Explain.**

8. The width of the Ochoa Community Pool is 20 feet. The length is twice as long as its width. What is the perimeter of the pool?

_____   _____

Lesson Check

1. What is the perimeter of a square window with sides 36 inches long?

   (A) 40 inches

   (B) 72 inches

   (C) 144 inches

   (D) 1,296 inches

2. What is the perimeter of the rectangle below?

   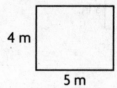

   4 m

   5 m

   (A) 11 meters   (C) 18 meters

   (B) 14 meters   (D) 400 meters

Spiral Review

3. Natalie drew the angle below.

   Which is the most reasonable estimate for the measure of the angle Natalie drew?
   (Lesson 11.2)

   (A) 30°

   (B) 90°

   (C) 180°

   (D) 210°

4. Ethan has 3 pounds of mixed nuts. How many ounces of mixed nuts does Ethan have? (Lesson 12.3)

   (A) 30 ounces

   (B) 36 ounces

   (C) 48 ounces

   (D) 54 ounces

5. How many lines of symmetry does the shape below appear to have? (Lesson 10.5)

   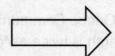

   (A) 0   (C) 2

   (B) 1   (D) more than 2

6. Which of the following comparisons is correct? (Lesson 9.7)

   (A) 0.70 > 7.0

   (B) 0.7 = 0.70

   (C) 0.7 < 0.70

   (D) 0.70 = 0.07

Area

Find the area of the rectangle or square.

1. 12 ft

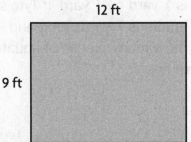

9 ft

$A = b \times h$

$= 12 \times 9$

108 square feet

2. 8 yd

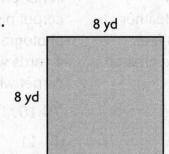

8 yd

_____

3. 15 m

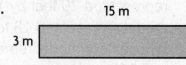

3 m

_____

4. 13 in.

6 in.

_____

5. 30 cm

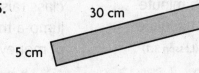

5 cm

_____

6. 14 ft

4 ft

_____

Problem Solving  REAL WORLD

7. Meghan is putting wallpaper on a wall that measures 8 feet by 12 feet. How much wallpaper does Meghan need to cover the wall?

_____

8. Bryson is laying down sod in his yard to grow a new lawn. Each piece of sod is a 1-foot by 1-foot square. How many pieces of sod will Bryson need to cover his yard if his yard measures 30 feet by 14 feet?

_____

Lesson Check

1. Ellie and Heather drew floor models of their living rooms. Ellie's model represented 20 feet by 15 feet. Heather's model represented 18 feet by 18 feet. Whose floor model represents the greater area? How much greater?

   (A) Ellie; 138 square feet

   (B) Heather; 24 square feet

   (C) Ellie; 300 square feet

   (D) Heather; 324 square feet

2. Tyra is laying down square carpet pieces in her photography studio. Each square carpet piece is 1 yard by 1 yard. If Tyra's photography studio is 7 yards long and 4 yards wide, how many pieces of square carpet will Tyra need?

   (A) 10

   (B) 11

   (C) 22

   (D) 28

Spiral Review

3. Typically, blood fully circulates through the human body 8 times each minute. How many times does blood circulate through the body in 1 hour? (Lesson 3.1)

   (A) 48   (C) 480

   (B) 240   (D) 4,800

4. Each of the 28 students in Romi's class raised at least $25 during the jump-a-thon. What is the least amount of money the class raised? (Lesson 3.5)

   (A) $5,200   (C) $660

   (B) $700   (D) $196

5. What is the perimeter of the shape below if 1 square is equal to 1 square foot?
   (Lesson 13.1)

   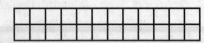

   (A) 12 feet

   (B) 14 feet

   (C) 24 feet

   (D) 28 feet

6. Ryan is making small meat loaves. Each small meat loaf uses $\frac{3}{4}$ pound of meat. How much meat does Ryan need to make 8 small meat loaves? (Lesson 8.4)

   (A) 4 pounds

   (B) 6 pounds

   (C) 8 pounds

   (D) $10\frac{2}{3}$ pounds

Area of Combined Rectangles

Find the area of the combined rectangles.

1.
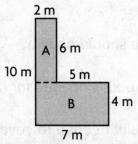
2 m

A 6 m

10 m   5 m

B   4 m

7 m

Area A = 2 × 6,
Area B = 7 × 4
12 + 28 = 40
40 square meters

2.
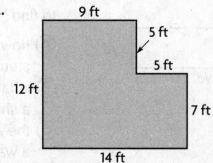
9 ft
5 ft
5 ft
12 ft
7 ft
14 ft

3.
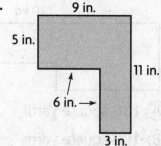
9 in.
5 in.
11 in.
6 in. →
3 in.

4.
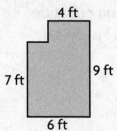
4 ft
7 ft   9 ft
6 ft

5.
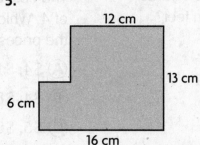
12 cm
13 cm
6 cm
16 cm

6.
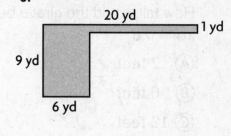
20 yd
1 yd
9 yd
6 yd
6 yd

Problem Solving REAL WORLD

Use the diagram for 7–8.

Nadia makes the diagram below to
represent the counter space she wants
to build in her craft room.

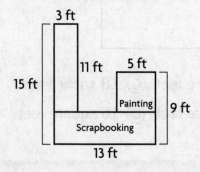
3 ft
11 ft   5 ft
15 ft
Painting   9 ft
Scrapbooking
13 ft

7. What is the area of the space that Nadia
has shown for scrapbooking?

_____

8. What is the area of the space she has
shown for painting?

_____

Lesson Check

1. What is the area of the combined rectangles below?

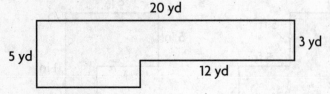

20 yd

5 yd

3 yd

12 yd

- (A) 136 square yards
- (B) 100 square yards
- (C) 76 square yards
- (D) 64 square yards

2. Marquis is redecorating his bedroom. What could Marquis use the area formula to find?

- (A) how much space should be in a storage box
- (B) what length of wood is needed for a shelf
- (C) the amount of paint needed to cover a wall
- (D) how much water will fill up his new aquarium

Spiral Review

3. Giraffes are the tallest land animals. A male giraffe can grow as tall as 6 yards. How tall would the giraffe be in feet? (Lesson 12.2)

- (A) 2 feet
- (B) 6 feet
- (C) 12 feet
- (D) 18 feet

4. Drew purchased 3 books for $24. The cost of each book was a multiple of 4. Which of the following could be the prices of the 3 books? (Lesson 5.4)

- (A) $4, $10, $10
- (B) $4, $8, $12
- (C) $5, $8, $11
- (D) $3, $7, $14

5. Esmeralda has a magnet in the shape of a square. Each side of the magnet is 3 inches long. What is the perimeter of her magnet? (Lesson 13.1)

- (A) 3 inches
- (B) 7 inches
- (C) 9 inches
- (D) 12 inches

6. What is the area of the rectangle below? (Lesson 13.2)

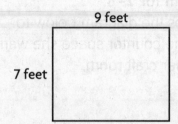

9 feet

7 feet

- (A) 63 square feet
- (C) 18 square feet
- (B) 32 square feet
- (D) 16 square feet

Find Unknown Measures

Find the unknown measure of the rectangle.

1. 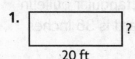 ?

20 ft

Perimeter = 54 feet

width = ____**7 feet**____

Think: $P = (2 \times l) + (2 \times w)$
$54 = (2 \times 20) + (2 \times w)$
$54 = 40 + (2 \times w)$
Since $54 = 40 + 14$, $2 \times w = 14$, and $w = 7$.

2.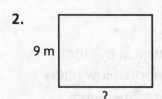

9 m

?

Perimeter = 42 meters

length = _____

3.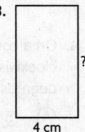

?

4 cm

Area = 28 square centimeters

height = _____

4.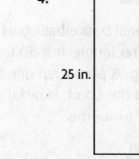

25 in.

Area = 200 square inches

base = _____

Problem Solving REAL WORLD

5. Susie is an organic vegetable grower. The perimeter of her rectangular vegetable garden is 72 yards. The width of the vegetable garden is 9 yards. How long is the vegetable garden?

6. An artist is creating a rectangular mural for the Northfield Community Center. The mural is 7 feet tall and has an area of 84 square feet. What is the length of the mural?

_____

_____

Lesson Check

1. The area of a rectangular photograph is 35 square inches. If the width of the photo is 5 inches, how tall is the photo?

   Ⓐ 5 inches

   Ⓑ 7 inches

   Ⓒ 25 inches

   Ⓓ 30 inches

2. Natalie used 112 inches of blue yarn as a border around her rectangular bulletin board. If the bulletin board is 36 inches wide, how long is it?

   Ⓐ 20 inches

   Ⓑ 38 inches

   Ⓒ 40 inches

   Ⓓ 76 inches

Spiral Review

3. A professional basketball court is in the shape of a rectangle. It is 50 feet wide and 94 feet long. A player ran one time around the edge of the court. How far did the player run? (Lesson 13.1)

   Ⓐ 144 feet

   Ⓑ 194 feet

   Ⓒ 238 feet

   Ⓓ 288 feet

4. On a compass, due east is a $\frac{1}{4}$ turn clockwise from due north. How many degrees are in a $\frac{1}{4}$ turn? (Lesson 11.2)

   Ⓐ 45°

   Ⓑ 60°

   Ⓒ 90°

   Ⓓ 180°

5. Hakeem's frog made three quick jumps. The first was 1 meter. The second jump was 85 centimeters. The third jump was 400 millimeters. What was the total length of the frog's three jumps? (Lesson 12.10)

   Ⓐ 189 centimeters

   Ⓑ 225 centimeters

   Ⓒ 486 millimeters

   Ⓓ 585 millimeters

6. Karen colors in squares on a grid. She colored $\frac{1}{8}$ of the squares blue and $\frac{5}{8}$ of the squares red. What fraction of the squares are not colored in? (Lesson 7.10)

   Ⓐ $\frac{1}{8}$

   Ⓑ $\frac{1}{4}$

   Ⓒ $\frac{1}{2}$

   Ⓓ $\frac{3}{4}$

Name _____

Problem Solving • Find the Area

Solve each problem.

1. A room has a wooden floor. There is a rug in the center of the floor. The diagram shows the room and the rug. How many square feet of the wood floor still shows?

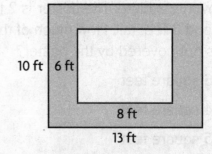

82 square feet
_____

Area of the floor: $13 \times 10 = 130$ square feet
Area of the rug: $8 \times 6 = 48$ square feet
Subtract to find the area of the floor still showing: $130 - 48 = 82$ square feet

2. A rectangular wall has a square window, as shown in the diagram.

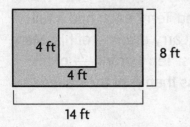

What is the area of the wall NOT including the window?

_____

3. Bob wants to put down new sod in his backyard, except for the part set aside for his flower garden. The diagram shows Bob's backyard and the flower garden.

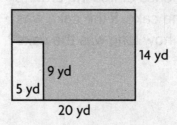

How much sod will Bob need?

_____

4. A rectangular painting is 24 inches wide and 20 inches tall without the frame. With the frame, it is 28 inches wide and 24 inches tall. What is the area of the frame not covered by the painting?

_____

5. One wall in Jeanne's bedroom is 13 feet long and 8 feet tall. There is a door 3 feet wide and 6 feet tall. She has a poster on the wall that is 2 feet wide and 3 feet tall. How much of the wall is visible?

_____

Lesson Check

1. One wall in Zoe's bedroom is 5 feet wide and 8 feet tall. Zoe puts up a poster of her favorite athlete. The poster is 2 feet wide and 3 feet tall. How much of the wall is not covered by the poster?

   (A) 16 square feet

   (B) 34 square feet

   (C) 35 square feet

   (D) 46 square feet

2. A garage door is 15 feet wide and 6 feet high. It is painted white, except for a rectangular panel 1 foot high and 9 feet wide that is brown. How much of the garage door is white?

   (A) 22 square feet

   (B) 70 square feet

   (C) 80 square feet

   (D) 81 square feet

Spiral Review

3. Kate baked a rectangular cake for a party. She used 42 inches of frosting around the edges of the cake. If the cake was 9 inches wide, how long was the cake?

   (Lesson 13.4)

   (A) 5 inches

   (B) 12 inches

   (C) 24 inches

   (D) 33 inches

4. Larry, Mary, and Terry each had a full glass of juice. Larry drank $\frac{3}{4}$ of his. Mary drank $\frac{3}{8}$ of hers. Terry drank $\frac{7}{10}$ of his. Who drank less than $\frac{1}{2}$ of their juice?

   (Lesson 6.6)

   (A) Larry

   (B) Mary

   (C) Mary and Terry

   (D) Larry and Terry

5. Which of the following statements is NOT true about the numbers 7 and 9?

   (Lesson 5.5)

   (A) 7 is a prime number.

   (B) 9 is a composite number.

   (C) 7 and 9 have no common factors other than 1.

   (D) 27 is a common multiple of 7 and 9.

6. Tom and some friends went to a movie. The show started at 2:30 P.M. and ended at 4:15 P.M. How long did the movie last?

   (Lesson 12.9)

   (A) 1 hour 35 minutes

   (B) 1 hour 45 minutes

   (C) 1 hour 55 minutes

   (D) 2 hours 15 minutes

Name _____

Chapter 13 Extra Practice

Lesson 13.1

Find the perimeter of the rectangle or square.

1. 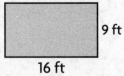 9 ft
16 ft

_____

2. 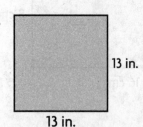 13 in.
13 in.

_____

3. 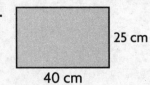 25 cm
40 cm

_____

4. 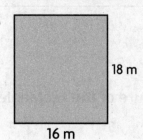 18 m
16 m

_____

Lesson 13.2

Find the area of the rectangle or square.

1. 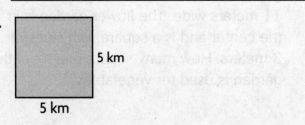 15 in.
12 in.

_____

2. 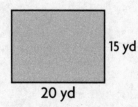 15 yd
20 yd

_____

3. 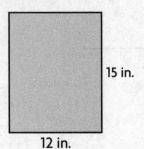 5 km
5 km

_____

4. 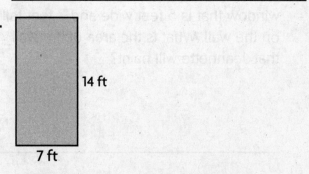 14 ft
7 ft

_____

Lesson 13.3

Find the area of the combined rectangles.

1.

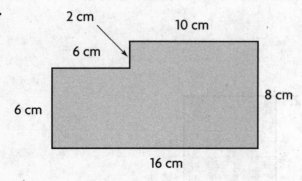

2.

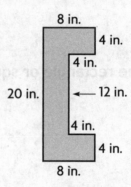

_____   _____

Lesson 13.4

Find the unkown measure of the rectangle.

1.

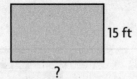

15 ft

?

Perimeter = 80 feet

base = _____

2.

?

7 mi

Area = 56 square miles

height = _____

Lesson 13.5

Solve.

1. Jeanette is painting a rectangular wall that is 10 feet long and 8 feet tall. There is a window that is 5 feet wide and 3 feet tall on the wall. What is the area of the wall that Jeannette will paint?

2. Rob has a combined flower and vegetable garden that is 9 meters long and 11 meters wide. The flower garden is in the center and is a square with sides of 3 meters. How many square meters of the garden is used for vegetables?

_____   _____

Name _____

Add Dollars and Cents

Essential Question How can you find sums of decimal amounts in dollars and cents?

UNLOCK the Problem REAL WORLD

Carlos bought a new skateboard for $99.46 and a helmet and pads for $73.49. How much did Carlos spend in all?

You add money amounts in the same way as you add whole numbers. Use the decimal point to line up the digits.

- What operation can you use to find the amount Carlos spent?

_____

 Use place value.

Add. $99.46 + $73.49

STEP 1	STEP 2	STEPS 3 and 4	STEP 5
Add the pennies. Regroup 15 pennies.	Add the dimes.	Add the ones. Add the tens.	Insert the decimal point and dollar sign.
$\begin{array}{r} 1 \\ \$\ 99.46 \\ +\ \$\ 73.49 \\ \hline 5 \end{array}$	$\begin{array}{r} 1 \\ \$\ 99.46 \\ +\ \$\ 73.49 \\ \hline 95 \end{array}$	$\begin{array}{r} 1\ \ 1 \\ \$\ 99.46 \\ +\ \$\ 73.49 \\ \hline 172\ 95 \end{array}$	$\begin{array}{r} 1\ \ 1 \\ \$\ 99.46 \\ +\ \$\ 73.49 \\ \hline \$172.95 \end{array}$

So, Carlos spent $172.95.

Try This! Find the sum.

A.

	$	2	3	.	1	8	
+	$	5	7	.	4	5	

B.

	$	1	9	.	0	7	
+	$	6	5	.	2	8	

Math Talk **Explain** how you know when to regroup.

Share and Show

1. Explain what is happening in Step 2.

_____

_____

_____

STEPS 1 and 2	STEPS 3 and 4	STEP 5
1	1 1	1 1
$84.60	$84.60	$84.60
+ $35.70	+ $35.70	+ $35.70
30	120 30	$120.30

Find the sum.

2. $3.09
    + $8.92

3. $26.08
    + $41.39

4. $ 7.26
    + $26.43

5. $30.47
    + $28.56

On Your Own

Find the sum.

6. $9.57
    + $4.09

7. $89.36
    + $ 3.85

8. $23.75
    + $10.98

9. $ 8.52
    + $36.07

10. $48.92
    + $ 7.08

11. $60.45
    + $17.42

12. $58.02
    + $73.54

13. $61.74
    + $60.57

Problem Solving REAL WORLD

14. Lena bought new inline skates for $49.99. The sales tax was $4.13. How much did Lena spend in all for her new inline skates?

_____

Name _____

Subtract Dollars and Cents

Essential Question How can you find differences between decimal amounts in dollars and cents?

UNLOCK the Problem · REAL WORLD

Sandi wanted to buy a new coat online. She figured out that the cost of the coat, with shipping, would be $84.24. The next week, Sandi bought the same coat in a local store on sale for a total of $52.47. How much did Sandi save by buying the coat on sale?

You subtract money amounts in the same way as you subtract whole numbers.

- Underline the information you need to solve the problem.

- What operation can you use to find the difference between the two prices?

_____

Use place value.

Subtract. $84.24 − $52.47

Use the decimal point to line up the digits. Work from right to left. Check each place to see if you need to regroup to subtract.

STEP 1	STEP 2	STEPS 3 and 4	STEP 5
Regroup 2 dimes and 4 pennies as 1 dime and 14 pennies. Subtract the pennies.	Regroup 4 dollars and 1 dime as 3 dollars and 11 dimes. Subtract the dimes.	Subtract the ones. Subtract the tens.	Insert the decimal point and dollar sign.
$$\begin{array}{r} {\scriptstyle 1\,14} \\ \$84.2\!\!\not{2}\!\!\not{4} \\ -\ \$52.47 \\ \hline 7 \end{array}$$	$$\begin{array}{r} {\scriptstyle 11} \\ {\scriptstyle 3\,\not{1}\,14} \\ \$\not{8}4.2\!\!\not{4} \\ -\ \$52.47 \\ \hline 77 \end{array}$$	$$\begin{array}{r} {\scriptstyle 11} \\ {\scriptstyle 3\,\not{1}\,14} \\ \$\not{8}4.2\!\!\not{4} \\ -\ \$52.47 \\ \hline 31\ 77 \end{array}$$	$$\begin{array}{r} {\scriptstyle 11} \\ {\scriptstyle 3\,\not{1}\,14} \\ \$\not{8}4.2\!\!\not{4} \\ -\ \$52.47 \\ \hline \$31.77 \end{array}$$

So, Sandi saved $31.77.

Math Talk **Explain** how you know in which places to regroup to subtract.

Share and Show

1. Find the difference. Regroup as needed.

$$\begin{array}{r} \$\ 7.14 \\ \underline{\$\ 4.38} \end{array}$$

Find the difference.

2.
$$\begin{array}{r} \$5.89 \\ -\ \$3.16 \\ \hline \end{array}$$

3.
$$\begin{array}{r} \$30.07 \\ -\ \$11.32 \\ \hline \end{array}$$

4.
$$\begin{array}{r} \$60.00 \\ -\ \$42.75 \\ \hline \end{array}$$

5.
$$\begin{array}{r} \$99.08 \\ -\ \$91.36 \\ \hline \end{array}$$

On Your Own

Find the difference.

6.
$$\begin{array}{r} \$9.08 \\ -\ \$7.26 \\ \hline \end{array}$$

7.
$$\begin{array}{r} \$73.45 \\ -\ \$12.13 \\ \hline \end{array}$$

8.
$$\begin{array}{r} \$90.00 \\ -\ \$42.17 \\ \hline \end{array}$$

9.
$$\begin{array}{r} \$80.03 \\ -\ \$49.53 \\ \hline \end{array}$$

10.
$$\begin{array}{r} \$15.36 \\ -\ \$\ 2.73 \\ \hline \end{array}$$

11.
$$\begin{array}{r} \$84.00 \\ -\ \$27.85 \\ \hline \end{array}$$

12.
$$\begin{array}{r} \$74.19 \\ -\ \$\ 8.46 \\ \hline \end{array}$$

13.
$$\begin{array}{r} \$79.62 \\ -\ \$23.58 \\ \hline \end{array}$$

Problem Solving

14. Bert earned $78.70 last week. This week he earned $93.00. How much more did he earn this week than last week?

_____

Name _____

Order of Operations

Essential Question How can you use the order of operations to find the value of expressions?

🔓 UNLOCK the Problem · REAL WORLD

At a visit to the Book Fair, Jana buys 7 hardcover books and 5 paperback books. She is going to give an equal number of books to each of her three cousins. How many books will each of Jana's cousins get?

To find the value of an expression involving parentheses, you can use the order of operations. Remember, the order of operations is a special set of rules that give you the order in which calculations are done in an expression.

First, perform operations inside the parentheses.

Then, multiply and divide from left to right.

Finally, add and subtract from left to right.

- What operation can you use to find the total number of books that Jana buys?

_____

- What operation can you use to find how many books each of Jana's cousins gets?

_____

🔑 Use the order of operations to find the value of $(7 + 5) \div 3$.

STEP 1

Perform operations in parentheses.

$(7 + 5) \div 3$

_____ $\div 3$

STEP 2

Use the order of operations. In this case, divide.

$12 \div 3$

So, each of Jana's cousins will get 4 books.

- **What if** Jana decides to keep 3 books for herself? How will this change the expression? How many books will each of Jana's cousins get?

_____

_____

_____

Math Talk What operation should you do first to find the values of $(6 + 2) \times 3$ and $6 + (2 \times 3)$? What is the value of each expression?

Share and Show

Write *correct* if the operations are listed in the correct order.
If not correct, write the correct order of operations.

1. $(4 + 5) \times 2$ multiply, add

_____

2. $8 \div (4 \times 2)$ multiply, divide

_____

3. $12 + (16 \div 4)$ add, divide

_____

4. $9 + 2 \times (3 - 1)$ add, multiply, subtract

_____

Follow the order of operations to find the value of the expression.
Show each step.

5. $6 + (2 \times 5)$

_____

6. $18 - (12 \div 4)$

_____

7. $8 \times (9 - 3)$

_____

8. $(12 + 8) \div 2 \times 3$

_____

On Your Own

Follow the order of operations to find the value of the expression.
Show each step.

9. $6 + (9 \div 3)$

_____

10. $(3 \times 6) \div 2$

_____

11. $(49 \div 7) + 5$

_____

12. $9 \times (8 - 2)$

_____

13. $45 \div (17 - 2)$

_____

14. $(32 + 4) \div 9 - 2$

_____

15. $8 \times 9 - (12 - 8)$

_____

16. $(36 - 4) + 8 \div 4$

_____

Problem Solving REAL WORLD

17. Mr. Randall bought 4 shirts, which were on sale. The shirts were
originally priced $20. The sales price of the shirts was $5 less than
the original price. Write and find the value of an expression for the
total amount that Mr. Randall paid for the shirts.

_____

Name _____

Divide by Multiples of Ten

Essential Question How can you use patterns to divide by multiples of ten?

🔓 UNLOCK the Problem REAL WORLD

A charity asked 10 volunteers to hand out 2,000 flyers about a fund-raising event. Each volunteer will get the same number of flyers. How many flyers will each volunteer hand out?

You can use patterns and a basic fact to divide by multiples of ten.

🔑 Example 1 Find $2,000 \div 10$.

Think: I know that $2 \div 1 = 2$, so $20 \div 10 = 2$.

$$20 \div 10 = 2$$
$$200 \div 10 = 20$$
$$2,000 \div 10 = 200$$

So, each volunteer will hand out _____ flyers.

Describe the pattern used to divide 2,000 by 10.

_____

_____

_____

🔑 Example 2 Find $2,800 \div 40$.

$28 \div 4 = 7$, so $280 \div 40 = $ _____.

$2,800 \div 40 = $ _____

Math Talk **Explain** how you can use basic facts to help divide by multiples of ten.

Share and Show

1. Find 6,000 ÷ 20.

Think: I can use patterns to divide, starting with 60 ÷ 20.

$6 ÷ 2 =$ _____, so $60 ÷ 20 =$ _____.

$600 ÷ 20 =$ _____

$6,000 ÷ 20 =$ _____

Divide. Use a pattern to help.

2. $8,000 ÷ 20 =$ _____

3. $4,000 ÷ 40 =$ _____

4. $1,200 ÷ 60 =$ _____

On Your Own

Divide. Use a pattern to help.

5. $9,000 ÷ 30 =$ _____

6. $5,000 ÷ 50 =$ _____

7. $1,800 ÷ 60 =$ _____

8. $7,000 ÷ 10 =$ _____

9. $3,200 ÷ 80 =$ _____

10. $6,300 ÷ 90 =$ _____

Problem Solving REAL WORLD

11. A group of musicians wants to sell a total of 1,000 tickets for 20 concerts. Suppose they sell the same number of tickets for each concert. How many tickets will they sell for each concert? **Explain** how you solved the problem.

_____

_____

_____

Name _____

Model Division with 2-Digit Divisors

Essential Question How can you use models to divide?

CONNECT You have used base-ten blocks to divide whole numbers by 1-digit divisors. You can follow the same steps to divide whole numbers by 2-digit divisors.

 UNLOCK the Problem REAL WORLD

🔑 **Activity** **Materials** ■ base-ten blocks

There are 154 children participating in a soccer tournament. There are 11 equal-sized teams of children. How many children are on each team?

- What do you need to find?

  _____

  _____

- What is the dividend? the divisor?

  _____

STEP 1

Use base-ten blocks to model 154 children. Show 154 as 1 hundred 5 tens 4 ones. Draw 11 ovals for the teams.

STEP 2

Share the base-ten blocks equally among 11 groups. Since there are not enough hundreds to share equally, regroup 1 hundred as 10 tens. There are now 15 tens. Share the tens and draw a vertical line segment for each ten.

STEP 3

If there are any tens left over, regroup each as 10 ones. Share the ones equally among 11 groups. Draw a small circle for each one.

◯◯◯◯◯◯◯◯◯◯◯

There are _____ ten(s) and _____ one(s) in each group.

So, there are _____ children on each team.

- Explain why you need to regroup in Step 3.

  _____

  _____

  _____

Math Talk **Explain** how you can check your answer.

Share and Show

1. Use base-ten blocks to find 182 ÷ 14. **Describe** the steps you took to find your answer.

_____

_____

_____

_____

Use base-ten blocks to divide.

2. 60 ÷ 12 = _____

3. 135 ÷ 15 = _____

On Your Own ·

Use base-ten blocks to divide.

4. 180 ÷ 10 = _____

5. 150 ÷ 15 = _____

6. 88 ÷ 11 = _____

7. 96 ÷ 16 = _____

8. 176 ÷ 11 = _____

9. 156 ÷ 13 = _____

Problem Solving REAL WORLD

10. Nicole has $250 in ten-dollar bills. How many ten-dollar bills does Nicole have?

_____

11. At Dante's party, 16 children share 192 crayons. At Maria's party, 13 children share 234 crayons. Each party splits the crayons up equally among the children attending. How many more crayons does each child at Maria's party get than each child at Dante's party? **Explain.**

_____

_____

Name _____

 **Checkpoint**

Concepts and Skills

Find the sum or difference. (pp. P259–P262)

1.	2.	3.	4.
$2.87 + $8.09	$7.65 − $5.23	$37.05 + $14.95	$30.00 − $12.69

Use base-ten blocks to divide. (pp. P267–P268)

5. 143 ÷ 11   6. 224 ÷ 16   7. 108 ÷ 18

_____   _____   _____

Follow the order of operations to find the value of the expression. Show each step. (pp. P263–P264)

8. $(8 \times 2) + 4$ 9. $16 - (3 \times 5)$ 10. $24 \div (15 - 7)$ 11. $15 \div (9 - 4) \times 4$

_____   _____

Divide. Use a pattern to help. (pp. P265–P266)

12. 6,000 ÷ 30   13. 2,000 ÷ 20   14. 3,200 ÷ 40   15. 8,100 ÷ 90

_____   _____   _____   _____

Problem Solving REAL WORLD

16. Ellis bought groceries that were worth $99.86. After using coupons, the bill was $84.92. How much did Ellis save by using coupons? (pp. P261–P262)

_____

Fill in the bubble completely to show your answer.

17. Taby buys a dog leash for $18.50 and a dog collar for $12.75. What is the total cost of the leash and the collar? (pp. P259–P260)

Ⓐ $5.75

Ⓑ $6.25

Ⓒ $30.25

Ⓓ $31.25

18. Mr. Martin pays $35.93 for shoes for himself and $18.67 for shoes for his son. How much more do Mr. Martin's shoes cost than his son's? (pp. P261–P262)

Ⓐ $17.26

Ⓑ $17.36

Ⓒ $23.24

Ⓓ $54.60

19. Chris and Susan each collect baseball cards. Chris has 75 cards and Susan has 93 cards. They want to combine their collections and divide the cards evenly between them. Which expression can they use to find the number of cards each of them should have? (pp. P263–P264)

Ⓐ 75 + 93 ÷ 2

Ⓑ 75 + (93 ÷ 2)

Ⓒ (75 + 93) × 2

Ⓓ (75 + 93) ÷ 2

20. A store expects 4,000 customers during its 20-hour sale. Suppose the same number of customers arrives each hour. How many customers come each hour? (pp. P265–P266)

Ⓐ 20

Ⓑ 200

Ⓒ 2,000

Ⓓ 8,000

Place Value Through Millions

Essential Question How can you read, write, and represent whole numbers through millions?

> ## ⚷ UNLOCK the Problem REAL WORLD

The population of Idaho is about 1,550,000. Write 1,550,000 in standard form, word form, and expanded form.

You know how to read and write numbers through hundred thousands. The place-value chart can be expanded to help you read and write greater numbers, like 1,550,000.

One million is 1,000 thousands and is written as 1,000,000. The millions period is to the left of the thousands period on a place-value chart.

- What is the value of the ten thousands place?

_____

PERIODS

MILLIONS			THOUSANDS			ONES		
Hundreds	Tens	Ones	Hundreds	Tens	Ones	Hundreds	Tens	Ones
		1,	5	5	0,	0	0	0
		1 × 1,000,000	5 × 100,000	5 × 10,000	0 × 1,000	0 × 100	0 × 10	0 × 1
		1,000,000	500,000	50,000	0	0	0	0

The place value of the 1 in 1,550,000 is millions.

Standard form: 1,550,000

Word Form: One million, five hundred fifty thousand

Expanded Form: 1,000,000 + 500,000 + 50,000

> **Math Talk** **Explain** how 8,000,000 is different than 800,000.

Try This! Use place value to read and write the number.

Standard Form: _____

Word Form: Sixty-two million, eighty thousand, one hundred twenty-six

Expanded Form: 60,000,000 + _____ +

80,000 + _____ + 20 + 6

Share and Show

1. Write the number 3,298,076 in word form and expanded form.

   Word Form: _____

   _____

   Expanded Form: _____

Read and write the number in two other forms.

2. fifty million, three thousand, eighty-seven

   _____

   _____

   _____

3. 60,000,000 + 400,000 + 200 + 30 + 9

   _____

   _____

   _____

On Your Own

Read and write the number in two other forms.

4. 70,000,000 + 8,000,000 + 20,000 + 8

   _____

   _____

   _____

5. twenty million, eleven thousand, twelve

   _____

   _____

   _____

Write the value of the underlined digit.

6. 3,3<u>5</u>6,000

   _____

7. 45,687,<u>9</u>09

   _____

8. <u>7</u>0,000,044

   _____

9. <u>3</u>0,051,218

   _____

Problem Solving

10. According to one organization, there are about 93,600,000 pet cats and about 77,500,000 pet dogs in the U.S. Are there more pet cats or pet dogs? **Explain** how you know.

    _____

    _____

    _____

    _____

Decimals and Place Value

Essential Question How can you use place value to read, write, and represent decimals?

CONNECT Decimals, like whole numbers, can be written in standard form, word form, and expanded form.

 UNLOCK the Problem REAL WORLD

One of the world's tiniest frogs lives in Asia. Adult males range in length from about 1.06 to 1.28 centimeters, about the size of a pea.

You can use a place-value chart to help you understand decimals. Whole numbers are to the left of the decimal point in the place-value chart, and decimal amounts are to the right of the decimal point. The value of each place is one-tenth of the place to its left.

- What decimals do you see in the problem?

  _____

- The numbers 1.06 and 1.28 are between which two whole numbers?

  _____

Use a place-value chart.

Write each of the decimals on a place-value chart. Be sure to line up each place and the decimal point.

ONES		TENTHS	HUNDREDTHS
1	.	0	6
1	.	2	8

The place-value position of the digit 8 in 1.28 is hundredths. The value of the digit 8 in 1.28 is 8 hundredths, or $8 \times \frac{1}{100}$ or 0.08.

You can also write 1.28 in word form and expanded form.

Word form: one and twenty-eight hundredths

Expanded form: $1 + 0.2 + 0.08$

Math Talk **Explain** why 1.28 is not one and twenty-eight tenths in word form.

Try This! Use place value to read and write the decimal.

Standard Form: _____

Word Form: three and forty-six hundredths

Expanded Form: $3 + $ _____ $ + $ _____

Share and Show

1. Write the decimal 4.06 in word form and expanded form.

   Word Form: _____

   Expanded Form: _____

Read and write the decimal in two other forms.

2. five and two tenths

   _____

3. $6 + 0.8 + 0.09$

   _____

   _____

On Your Own

Read and write the decimal in two other forms.

4. seven and three hundredths:

   _____

5. $2 + 0.3 + 0.01$

   _____

Write the value of the underlined digit.

6. 4.<u>5</u>6

   _____

7. 5.0<u>9</u>

   _____

8. <u>7</u>.4

   _____

9. 1.3<u>2</u>

   _____

Problem Solving

10. James is 1.63 meters tall. Write James's height in word form. **Explain** how you found your answer.

    _____

    _____

    _____

11. Ani was told to write the number four and eight hundredths. She wrote 4.8. **Explain** whether or not you think Ani is correct. If you think she is not correct, write the number correctly.

    _____

    _____

Name _____

Round Decimals

Essential Question How can you round decimal amounts, including amounts of money, to the nearest whole number or dollar?

UNLOCK the Problem REAL WORLD

Ami sells fruits and nuts at an outdoor market. She sold a bag of nuts that weighed 1.35 pounds. About how much did the bag of nuts weigh, rounded to the nearest whole number?

- Underline the information that you need to find.

You know that you can use a number line or place value to round whole numbers. You can use the same strategies to round decimals.

🔑 Use a number line.

To round a decimal to the nearest whole number, find the whole numbers it is between.

_____ < 1.35 < _____

Use a number line to see which whole number 1.35 is closer to.

1.35 is closer to _____ than _____.

So, the bag of nuts weighed about _____ pound.

Math Talk **Explain** how rounding decimals is like rounding whole numbers.

1. **What if** Ami sold a bag of nuts that weighed 2.82 pounds? About how much does the bag weigh, rounded to the nearest whole number?

_____

2. **Describe** how you would round $3.90 to the nearest whole dollar.

_____

_____

_____

Share and Show

1. Round $2.67 to the nearest dollar. Locate and mark $2.67 on the

   number line. Which whole dollar is it closest to? _____

   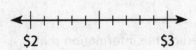

Round to the nearest dollar or to the nearest whole number.

2. $0.78	3. 2.1	4. 3.5	5. $4.50
_____	_____	_____	_____

On Your Own ....

Round to the nearest dollar or to the nearest whole number.

6. $1.70	7. 2.2	8. $3.99	9. 3.45
_____	_____	_____	_____

10. $1.53	11. 0.9	12. $0.19	13. 4.38
_____	_____	_____	_____

Problem Solving REAL WORLD

14. Candice spent $13.55 at the arts and crafts fair. How much
    money did Candice spend, rounded to the nearest dollar?

    _____

15. Mr. Marsh bought 2.25 pounds of American cheese. About
    how many pounds of cheese did Mr. Marsh buy?

    _____

Name _____

Place Value to Compare Decimals

Essential Question How can you use place value to compare decimals?

 **UNLOCK the Problem** REAL WORLD

Hummingbirds are small, fast, light birds that feed on flowers, trees, and insects. Suppose a particular hummingbird weighs 0.16 ounces. A nickel weighs about 0.18 ounces. Does the hummingbird weigh more or less than a nickel?

- What do you need to do to solve the problem?

_____

- Circle the numbers you need to compare.

Use a place-value chart.

Write each of the decimals on a place-value chart. Be sure to line up each place and the decimal point. Then compare the numbers in each place.

Ones		Tenths	Hundredths
0	.	1	6
0	.	1	8

$0 = 0$ $1 =$ _____ $6 <$ _____

Since 6 ◯ 8, 0.16 ◯ 0.18, and 0.18 ◯ 0.16

So, the hummingbird weighs _____ a nickel.

Math Talk **Explain** why you start comparing the decimals by comparing the ones place.

Try This! Use a place-value chart to compare the decimals.

Write <, >, or =.

A. 1.32 ◯ 1.34

B. 0.67 ◯ 0.6

C. 0.99 ◯ 0.99

Share and Show

1. Use the place-value chart below to compare the decimals.
   Write <, >, or =.

Ones		Tenths	Hundredths
3	.	0	5
3	.	0	1

   3 = 3   0 = _____   5 ◯ 1

   So, 3.05 ◯ 3.01.

Compare the decimals. Write <, >, =.

2. 7.24 ◯ 7.42 | 3. 8.80 ◯ 8.81 | 4. 0.11 ◯ 0.11 | 5. 4.33 ◯ 4.31

On Your Own

Compare the decimals. Write <, >, =.

6. 0.04 ◯ 0.04 | 7. 1.1 ◯ 1.7 | 8. 0.34 ◯ 0.36 | 9. 4.04 ◯ 4.01

10. 9.67 ◯ 9.63 | 11. 1.4 ◯ 1.42 | 12. 0.02 ◯ 0.2 | 13. 5.4 ◯ 5.40

Use a place-value chart to order the decimals from least to greatest.

14. 0.59, 0.51, 0.52

15. 7.15, 7.18, 7.1

16. 1.3, 1.33, 1.03

_____ _____ _____

Problem Solving REAL WORLD

17. Jill, Ally, and Maria ran the 50-yard dash. Jill ran the race in
    6.87 seconds. Ally ran the race in 6.82 seconds. Maria ran the
    race in 6.93. Who ran the race the fastest? **Explain** how you
    can use a place-value chart to find the answer.

    _____

    _____

    _____

Name _____

Decompose Multiples of 10, 100, 1,000

Essential Question How can you find factors of multiples of 10, 100, and 1,000?

🔑 UNLOCK the Problem REAL WORLD

Architects make scale models of buildings before they build the real thing. The height of an actual building is going to be 1,200 feet. The scale model is 12 feet tall. How many times the height of the model is the height of the actual building?

You can decompose a multiple of 10, 100, or 1,000 by finding factors.

- What do you need to find?

_____

- Circle the numbers you need to use to solve the problem.

🔒 One Way Use mental math and a pattern.

Decompose 1,200.

$1,200 = $ _____ $\times 1$

$1,200 = $ _____ $\times 10$

$1,200 = $ _____ $\times 100$

So, the building is 100 times the height of the model.

🔒 Another Way Use place value.

Decompose 1,200.

$1,200 = 12$ hundreds $= 12 \times$ _____

So, $1,200 = 12 \times 100$.

Remember

A multiple of 10, 100, or 1,000 is a number that has a factor of 10, 100, or 1,000.

Math Talk **Explain** the difference between factors and multiples.

- Explain how you use mental math and a pattern to find factors of multiples of 10, 100, or 1,000.

_____

_____

_____

_____

Share and Show

1. Complete the exercise below to decompose 2,800.

   2,800 = _____ × 1

   2,800 = _____ × 10

   2,800 = _____ × 100

2. Complete the exercise below to decompose 930.

   930 = _____ tens = _____ × _____

Decompose each number.

3. 80 = _____ | 4. 320 = _____ | 5. 8,000 = _____

On Your Own ·

Decompose each number.

6. 90 = _____ | 7. 40 = _____ | 8. 890 = _____

9. 300 = _____ | 10. 7,000 = _____ | 11. 3,700 = _____

Correct the error. Write the correct decomposition.

12. 560 = 56 × 100

_____

13. 4,300 = 43 × 1,000

_____

14. 6,000 = 60 × 10

_____

Problem Solving

15. Jon goes to the bank with $990. How many ten-dollar bills can he get? Show how you found your answer.

_____

_____

Name _____

Number Patterns

Essential Question How can you use multiplication to describe a pattern?

 UNLOCK the Problem REAL WORLD

You know how to use a rule and a first term to write a sequence. Now, you will describe a sequence using a rule.

🔑 **Describe a pattern.**

A scientist counts the number of lily pads in a pond each day. She records the number of lily pads in the table below. How many lily pads will be in the pond on days 5 and 6?

Day	1	2	3	4
Lilly Pads	8	16	32	64

- Do the numbers in the sequence increase or decrease?

  _____

- Underline the information you need to find.

STEP 1 Describe the sequence.

THINK: How do I get from one term to the next?

Try multiplying by 2 since $8 \times 2 = 16$.

8, 16, 32, 64

Write a rule to describe the number of lily pads in the pond.

RULE: _____.

STEP 2 Find the next two terms in the sequence.

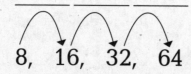

×2 ×2 ×2

8, 16, 32, 64, _____, _____

So, there will be _____ lily pads on day 5 and _____ lily pads on day 6.

> **Math Talk** **Explain** how you know the rule isn't add 8.

Share and Show

1. Find the next two numbers in the pattern below.

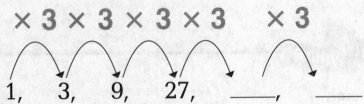

$\times 3 \times 3 \times 3 \times 3 \quad \times 3$

1, 3, 9, 27, _____, _____

Describe the pattern. Then find the next two numbers in the pattern.

2. 1, 2, 4, 8, _____, _____

_____

3. 7, 14, 28, 56, _____, _____

On Your Own

Describe the pattern. Then find the next two numbers in the pattern.

4. 1, 4, 16, 64, _____, _____

_____

5. 2, 6, 18, 54, _____, _____

_____

Determine the pattern and use it to fill in the blanks.

6. 1, 5, 25, _____, 625

7. 3, 6, _____, 24, _____

8. 2, _____, 32, _____, 512

Problem Solving REAL WORLD

9. A clothing store starts selling a new type of sneaker. The table shows the number of pairs of sneakers sold in the first four weeks. If the pattern continues, how many pairs of sneakers will the store sell in weeks 5 and 6? **Explain.**

Week	1	2	3	4
Pairs Sold	5	10	20	40

_____

_____

Name _____

✔ Checkpoint

Concepts and Skills

Round to the nearest whole dollar or to the nearest whole number. (pp. 275–276)

1. $7.23

2. 2.89

3. 0.52

4. $9.49

_____   _____   _____   _____

Compare the decimals. Write <, >, or =. (pp. P277–P278)

5. 0.6 ◯ 0.60   6. 5.08 ◯ 5.80   7. 8.14 ◯ 8.17   8. 7.37 ◯ 7.32

Read and write the numbers in two other forms. (pp. P271–P272)

9. seventy-five million, three hundred thousand, two hundred seven

_____

_____

_____

10. 30,000,000 + 40,000 + 6,000 + 20 + 2

_____

_____

_____

Decompose each number. (pp. P279–P280)

11. 20 = _____

12. 740 = _____

13. 6,000 = _____

Problem Solving REAL WORLD

14. A new music website is keeping track of the number of members that join. The table shows the number of members in the first four days. If the pattern continues, how many members will the website have on day 6? **Explain** how you found your answer. (pp. P281–P282)

Day	1	2	3	4
Members	5	15	45	135

_____

_____

_____

15. A particular female Asian elephant weighs 4.63 tons. What is this decimal written in word form? (pp. P273–P274)

(A) four and sixty-three tenths

(B) four and sixty-three hundredths

(C) four hundred and sixty-three

(D) four and sixty-three thousandths

16. Joe, Adam, Michael, and Carl all work at an office. Joe earns $15.53 per hour. Adam earns $15.59 per hour. Carl earns $15.95 per hour. Michael earns $15.91. Who earns the most money per hour? (pp. P277–P278)

(A) Joe

(B) Adam

(C) Carl

(D) Michael

17. Which number is ninety-eight million, forty thousand, six hundred fifty three written in another form? (pp. P271–P272)

(A) 98,040,653

(B) 98,400,653

(C) 98,046,053

(D) 98,40,653

18. Which rule describes the pattern below? (pp. P281–P282)

3, 12, 48, 192

(A) Multiply by 2.

(B) Multiply by 3.

(C) Add 9.

(D) Multiply by 4.

Name _____

Add Related Fractions

Essential Question How can you add fractions when one denominator is a multiple of the other?

When you add fractions, you find how many equal-size pieces there are in all. The denominator shows the size of the pieces. To add fractions with denominators that are not the same, first find equivalent fractions with the same denominator.

Activity

Materials ▪ fraction strips

Find $\frac{1}{2} + \frac{2}{6}$.

STEP 1 Model the problem.

Think: To add fractions, you need to count equal size pieces. The $\frac{1}{2}$ strip and the $\frac{1}{6}$ strip are different sizes.

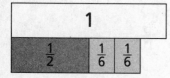

STEP 2 Show $\frac{1}{2}$ using $\frac{1}{6}$ strips.

$$\frac{1}{2} = \frac{}{6}$$

STEP 3 Add. Use the equivalent fraction you found. Find $\frac{3}{6} + \frac{2}{6}$.

How many $\frac{1}{6}$ strips are there?

_____

Write the sum. $\frac{3}{6} + \frac{2}{6} =$ _____

So, $\frac{1}{2} + \frac{2}{6} =$ _____.

- **Describe** how the sizes of the $\frac{1}{2}$ strip and the $\frac{1}{6}$ strip compare. Then describe how the denominators of the fractions $\frac{1}{2}$ and $\frac{1}{6}$ are related.

_____

_____

_____

_____

Math Talk **Explain** how you know $\frac{1}{2}$ and $\frac{3}{6}$ are equivalent fractions.

Share and Show

1. **Explain** which fraction strips you could use to add $\frac{1}{3}$ and $\frac{3}{6}$.

_____

_____

_____

2. Use fraction strips to add $\frac{1}{4} + \frac{2}{8}$.

$\frac{1}{4} + \frac{2}{8} =$ _____

Add. Use fraction strips to help.

3. $\frac{1}{4} + \frac{1}{2} =$ _____

4. $\frac{1}{2} + \frac{3}{8} =$ _____

5. $\frac{1}{2} + \frac{3}{10} =$ _____

On Your Own ....

Add. Use fraction strips to help.

6. $\frac{1}{3} + \frac{2}{6} =$ _____

7. $\frac{1}{5} + \frac{3}{10} =$ _____

8. $\frac{3}{8} + \frac{1}{4} =$ _____

9. $\frac{5}{12} + \frac{1}{3} =$ _____

10. $\frac{1}{3} + \frac{8}{12} =$ _____

11. $\frac{8}{10} + \frac{1}{5} =$ _____

Problem Solving REAL WORLD

12. Paola used $\frac{1}{4}$ of a carton of eggs today and $\frac{4}{12}$ of the carton yesterday. What fraction of the carton of eggs did she use in all? **Explain** how you found your answer.

_____

_____

Name _____

Subtract Related Fractions

Essential Question How can you subtract fractions when one denominator is a multiple of the other?

When you subtract fractions, you must use equal-size pieces.
To subtract fractions with different denominators, first find equivalent fractions with the same denominator. You can also compare to find the difference.

🔑 Activity

Materials ■ fraction strips

Find $\frac{5}{8} - \frac{1}{4}$.

🔑 One Way Find an equivalent fraction.

Model the problem.

Think: You need to subtract $\frac{1}{4}$ from $\frac{5}{8}$, but the $\frac{1}{4}$ strip and the $\frac{1}{8}$ strips are different sizes.

Show $\frac{1}{4}$ using $\frac{1}{8}$ strips.

$\frac{1}{4} = \frac{}{8}$

Subtract. Use the equivalent fraction you found.

Find $\frac{5}{8} - \frac{2}{8}$.

Write the difference. $\frac{5}{8} - \frac{2}{8} =$ _____

So, $\frac{5}{8} - \frac{1}{4} =$ _____.

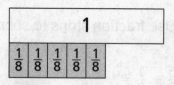

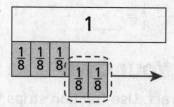

🔑 Another Way Compare to find the difference.

Model the problem.

Think: The $\frac{1}{4}$ strip is the same size as two $\frac{1}{8}$ strips.

Compare the $\frac{1}{4}$ strip to the five $\frac{1}{8}$ strips. Find the difference.

$\frac{5}{8} - \frac{1}{4} =$ _____.

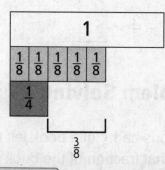

Math Talk **Explain** how the $\frac{1}{4}$ strip is related to the $\frac{1}{8}$ strip. Then describe how the denominators 4 and 8 are related.

© Houghton Mifflin Harcourt Publishing Company

Share and Show

1. A student subtracted $\frac{2}{3}$ from 1 whole as shown at the right. Explain the student's method. Then find the difference.

_____

_____

_____

_____

1		
$\frac{1}{3}$	$\frac{1}{3}$	$\frac{1}{3}$
$\frac{1}{3}$	$\frac{1}{3}$	

$\frac{1}{3}$

2. Use fraction strips to subtract $\frac{5}{6} - \frac{1}{2}$.

$\frac{5}{6} - \frac{1}{2} =$ _____

Subtract. Use fraction strips to help.

3. $\frac{1}{2} - \frac{3}{8} =$ _____

4. $1 - \frac{2}{5} =$ _____

5. $\frac{2}{4} - \frac{2}{12} =$ _____

On Your Own

Subtract. Use fraction strips to help.

6. $\frac{4}{5} - \frac{2}{10} =$ _____

7. $\frac{7}{8} - \frac{3}{4} =$ _____

8. $\frac{5}{6} - \frac{2}{3} =$ _____

9. $\frac{7}{10} - \frac{2}{5} =$ _____

10. $\frac{2}{6} - \frac{1}{3} =$ _____

11. $\frac{6}{8} - \frac{1}{2} =$ _____

Problem Solving REAL WORLD

12. Boris had $\frac{2}{3}$ of a book left to read. He read $\frac{1}{6}$ of the book today. What fraction of the book does he have left to read now? **Explain** how you found your answer.

_____

_____

Name _____

Compare Fraction Products

Essential Question How does the size of the product compare to the size of each factor when multiplying fractions in real-world situations?

🔑 UNLOCK the Problem   REAL WORLD

🔒 One Way Use a model.

A. Serena uses $\frac{2}{3}$ yard of fabric to make a pillow. How much fabric does she need to make 3 pillows?

- Shade the model to show 3 groups of $\frac{2}{3}$.

- Write an expression for three groups of $\frac{2}{3}$: _____ × _____.

- What can you say about the product when $\frac{2}{3}$ is multiplied by a whole number? Write *greater than* or *less than*. The product is _____ $\frac{2}{3}$.

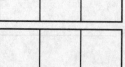

B. Serena has 3 yards of fabric. She uses $\frac{2}{3}$ of it to make a blanket. How much fabric does she use to make the blanket?

- There are 3 wholes. Each represents one yard.

- Shade $\frac{2}{3}$ of each whole.

- Write an expression for $\frac{2}{3}$ of three wholes: _____ × _____

- What can you say about the product when 3 is multiplied by a fraction less than 1? Write *greater than* or *less than*. The product is _____ 3.

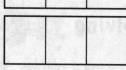

🔒 Another Way Use a number line.

A. Show $\frac{2}{3} \times 2$.

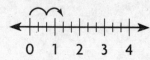

B. Show $\frac{2}{3} \times 3$.

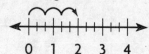

Complete each statement with *greater than* or *less than*.

- The product of $\frac{2}{3}$ and 2 is _____ $\frac{2}{3}$.

- The product of a whole number greater than 1 and $\frac{2}{3}$ will be _____ the whole number factor.

Math Talk What if a different fraction was multiplied by 2 and 3? Would your statements still be true? **Explain.**

Share and Show

1. Complete the statement with *greater than* or *less than*.

   $2 \times \frac{3}{4}$ will be _____ $\frac{3}{4}$.

   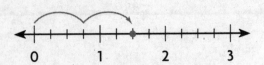

Complete each statement with *greater than* or *less than*.

2. $3 \times \frac{2}{5}$ will be _____ 3.

3. $3 \times \frac{1}{3}$ will be _____ $\frac{1}{3}$.

On Your Own

Complete each statement with *greater than* or *less than*.

4. $3 \times \frac{3}{8}$ will be _____ $\frac{3}{8}$.

5. $\frac{5}{6} \times 5$ will be _____ $\frac{5}{6}$.

6. $\frac{3}{10} \times 6$ will be _____ $\frac{3}{10}$.

7. $4 \times \frac{5}{9}$ will be _____ 4.

Problem Solving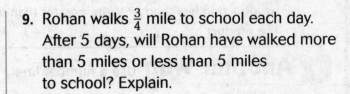

8. Celia wants to sew 4 pillows. She needs $\frac{3}{8}$ yard of fabric for each pillow. Will she need more than $\frac{3}{8}$ yard or less than $\frac{3}{8}$ yard of fabric to make all the pillows? Explain.

   _____

   _____

   _____

9. Rohan walks $\frac{3}{4}$ mile to school each day. After 5 days, will Rohan have walked more than 5 miles or less than 5 miles to school? Explain.

   _____

   _____

   _____

Repeated Subtraction with Fractions

Essential Question How can you use repeated subtraction to solve problems involving division with fractions?

🔑 UNLOCK the Problem REAL WORLD

Mr. Jones is making snacks for his family. He has 3 cups of almonds and is dividing them into $\frac{1}{2}$-cup portions. How many portions can he make?

You have used repeated subtraction to divide whole numbers. Now, you will use repeated subtraction to solve a problem involving division by a fraction.

🔑 **Use repeated subtraction to divide 3 by $\frac{1}{2}$.**

- What do you need to find?

  _____

  _____

- What other operation can you use instead of repeated subtraction to solve the problem?

  _____

STEP 1 Start at 3 and count back $\frac{1}{2}$.

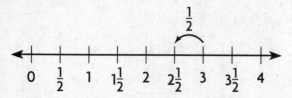

STEP 2 Subtract by $\frac{1}{2}$ until you reach 0 or get as close to it as possible.

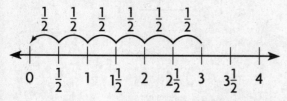

Math Talk **Explain** why you count the number of groups of $\frac{1}{2}$.

STEP 3 Find the number of times you counted back by $\frac{1}{2}$.

You counted _____ groups of $\frac{1}{2}$ to reach 0.

So, Mr. Jones can make _____ half-cup portions of almonds.

Share and Show

1. Use repeated subtraction and the number line to find $2 \div \frac{1}{4}$.

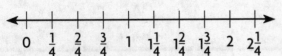

Start subtracting at _____.

Count back by groups of _____.

How many groups did you count to reach 0? _____

Use repeated subtraction to divide.

2. $2 \div \frac{1}{3}$

_____

3. $5 \div \frac{1}{2}$

_____

4. $1 \div \frac{1}{8}$

_____

On Your Own . . . .

Use repeated subtraction to divide.

5. $1 \div \frac{1}{5}$

_____

6. $2 \div \frac{1}{2}$

_____

7. $4 \div \frac{1}{3}$

_____

8. $2 \div \frac{1}{5}$

_____

9. $7 \div \frac{1}{2}$

_____

10. $3 \div \frac{1}{4}$

_____

Problem Solving REAL WORLD

11. You are putting raisins into snack bags. You have 3 cups of raisins. You want to put $\frac{1}{3}$ cup of raisins in each bag. How many bags can you make?

_____

12. Margaret is cutting straws that are 4 inches long into $\frac{1}{2}$-inch pieces. She has two straws. She needs twenty $\frac{1}{2}$-inch pieces. Does she have enough to cut 20 pieces? **Explain.**

_____

_____

Name _____

Fractions and Division

Essential Question How can you write division problems as fractions?

Division and fractions both show sharing equal numbers of things or making equal-size groups. You can write division problems as fractions.

🔓 UNLOCK the Problem REAL WORLD

Mavi and her 2 sisters want to share 4 small pizzas equally. How much pizza will each person have?

> • How many people want to share the pizzas?
> _____

Think: What is 4 divided by 3, or $4 \div 3$?

Each pizza is divided into _____ equal slices.

How many slices are in 4 pizzas? _____

What fraction of the pizza is each slice? _____

How many $\frac{1}{3}$-size slices does each sister get? _____

What fraction of the pizzas does each sister get? _____

So, $4 \div 3$ is the same as $\frac{4}{3}$.

Math Talk How can you write $\frac{4}{3}$ as a mixed number?

Share and Show

1. Alex baked a pan of corn bread and cut it into 12 equal-size pieces. Alex and his 3 sisters want to share the pieces equally.

   What division problem can you write to solve the problem? _____

   Write the division problem as a fraction. _____

Write the division problem as a fraction. Write each fraction greater than 1 as a whole number or mixed number.

2. $6 \div 2$	3. $1 \div 4$	4. $1 \div 3$	5. $32 \div 8$
_____	_____	_____	_____

On Your Own ....

Write the division problem as a fraction. Write each fraction greater than 1 as a whole number or mixed number.

6. $5 \div 6$	7. $3 \div 2$	8. $1 \div 8$	9. $2 \div 4$
_____	_____	_____	_____

10. $12 \div 3$	11. $9 \div 4$	12. $11 \div 2$	13. $8 \div 6$
_____	_____	_____	_____

Problem Solving REAL WORLD

14. Stefan and his 2 friends want to share 16 muffins equally. Will each friend get more than or less than 5 whole muffins? **Explain** how you know.

_____

_____

_____

Name _____

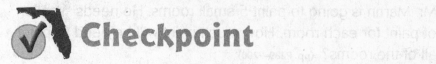 **Checkpoint**

Concepts and Skills

Complete each statement with *greater than* **or**
less than. **(pp. P289–P290)**

1. $3 \times \frac{3}{9}$ will be _____ 3.　　**2.** $\frac{7}{8} \times 3$ will be _____ $\frac{7}{8}$.

Add or subtract. Use fraction strips to help. **(pp. P285–P288)**

3. $\frac{1}{2} + \frac{2}{10} =$ _____

4. $\frac{1}{4} + \frac{5}{8} =$ _____

5. $\frac{4}{6} + \frac{1}{3} =$ _____

6. $1 - \frac{5}{6} =$ _____

7. $\frac{7}{8} - \frac{1}{4} =$ _____

8. $\frac{3}{5} - \frac{4}{10} =$ _____

**Write the division problem as a fraction. Write each fraction greater
than 1 as a whole number or mixed number.** **(pp. P293–P294)**

9. $7 \div 8 =$ _____

10. $8 \div 5 =$ _____

11. $16 \div 3 =$ _____

Use repeated subtraction to divide. **(pp. P291–P292)**

12. $3 \div \frac{1}{5} =$ _____

13. $4 \div \frac{1}{2} =$ _____

14. $6 \div \frac{1}{3} =$ _____

Problem Solving REAL WORLD

15. Manny had $\frac{3}{4}$ of his paper written. He wrote another $\frac{1}{8}$ of the paper
today. What fraction of the paper does he have left to write now?
Explain how you found your answer. **(pp. P285–P288)**

_____

_____

_____

Fill in the bubble completely to show your answer.

16. Mr. Martin is going to paint 5 small rooms. He needs $\frac{3}{4}$ gallon of paint for each room. How much paint will he need to paint all of the rooms? (pp. P289–P290)

  Ⓐ less than $\frac{3}{4}$ gallon

  Ⓑ more than $\frac{3}{4}$ gallon

  Ⓒ exactly $\frac{3}{4}$ gallon

  Ⓓ exactly 5 gallons

17. A chef is preparing individual-size pies. She has 4 cups of strawberries to put in the pies. She wants to put $\frac{1}{4}$ cup of strawberries in each pie. How many pies can she make?
(pp. P291–P292)

  Ⓐ 4

  Ⓑ 8

  Ⓒ 14

  Ⓓ 16

18. Which shows the division problem $6 \div 4$ written as a fraction or mixed number? (pp. P293–P294)

  Ⓐ $\frac{4}{6}$

  Ⓑ $1\frac{1}{4}$

  Ⓒ $1\frac{2}{4}$

  Ⓓ $2\frac{2}{4}$

19. Pablo ate $\frac{1}{4}$ of a pizza yesterday and $\frac{3}{8}$ of the pizza today. What fraction of the pizza did he eat in all? (pp. P285–P286)

  Ⓐ $\frac{5}{8}$

  Ⓑ $\frac{4}{12}$

  Ⓒ $\frac{4}{8}$

  Ⓓ $\frac{3}{8}$

Name _____

Locate Points on a Grid

Essential Question How can you use ordered pairs to locate points on a grid?

An **ordered pair** is a pair of numbers that names a point on a grid. The first number shows how many units to move horizontally. The second number shows how many units to move vertically.

(2 , 4)

Move 2 units right from 0.   Then move 4 units up.

🔑 UNLOCK the Problem REAL WORLD

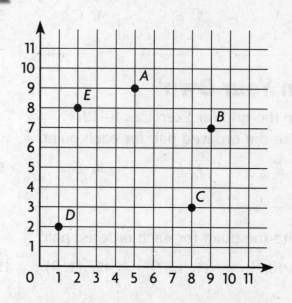

At the airport, passengers travel from one terminal to another in shuttle buses. The shuttle buses travel in a route that begins at Terminal A. Where is Terminal A?

Count units on the grid to find out.

• Start at zero.

• Move right 5 units.

• From there, move up 9 units.

Terminal A is located at (5, 9).

Try This!

What terminal is located at (8, 3)? Explain how you know.

_____

_____

_____

_____

Math Talk **Explain** why (3, 6) and (6, 3) are two different ordered pairs.

Share and Show

1. To graph the point (6, 3), where do you start? In which direction and how many units will you move first? What will you do next? Describe the steps and record them on the grid.

_____

_____

Use the grid for Exercises 2–5. Write the ordered pair for each point.

2. A	3. B	4. C	5. D
_____	_____	_____	_____

On Your Own

Use the grid for Exercises 6–13.
Write the ordered pair for each point.

6. E	7. F	8. G	9. H
_____	_____	_____	_____

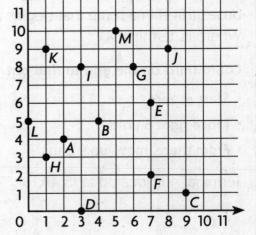

Write the point for each ordered pair.

10. (3, 8)	11. (8, 9)	12. (1, 9)	13. (0, 5)
_____	_____	_____	_____

Problem Solving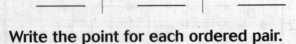

There are four photos on each page of a photo album. Complete the table. Write the data in the table as ordered pairs. Then graph the ordered pairs on the grid. Use the number of pages as the first number and the number of photos as the second number in the ordered pair.

14.

Number of Pages	1		3	4
Number of Photos	4	8		

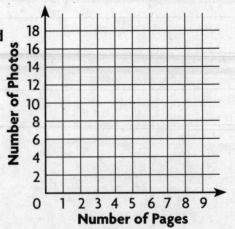

_____

Name _____

Area and Tiling

Essential Question How can you use tiling to find the area of a rectangle?

🔓 UNLOCK the Problem REAL WORLD

Rhonda is tiling the floor of her new sunroom. The diagram shows the layout of the tiles. Each tile measures 4 square feet. What is the area of Rhonda's sunroom floor?

To find the area of the sunroom floor, you can combine the areas of the half tiles and the whole tiles.

Find the area of the sunroom floor.

STEP 1 Find the area of the half tiles.

Count the number of half tiles. _____

1 tile = 4 square feet, so 1 half tile = 4 ÷ 2 or _____ square feet.

Multiply the number of half tiles by _____ square feet to find the area of the half tiles:

_____ × _____ = _____ square feet

STEP 2 Find the area of the whole tiles.

Find the number of whole tiles: $b \times h =$ _____ × _____ = _____ tiles

Since the area of 1 tile is _____ square feet, multiply the number of whole tiles by _____ to find the area of the whole tiles.

_____ × _____ = _____ square feet

STEP 3 Find the total area.

Add the areas of the half tiles and whole tiles.

half tiles   whole tiles
↓   ↓

_____ + _____ = _____ square feet

So, the area of Rhonda's sunroom floor is _____ square feet.

- Underline what you are asked to find.
- Circle the information you will use to solve the problem.

Rhonda's Sunroom Floor

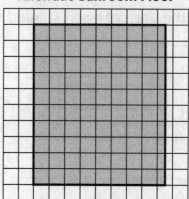

1 tile = 4 square feet

Remember

The formula for the area of a rectangle is $A = b \times h$ or $l \times w$.

Math Talk **Explain** how to find the area of 6 half tiles if 1 whole tile is 9 square inches.

Share and Show

1. Find the area of the shaded shape.

    STEP 1 Find the area of the half squares:

    _____ half squares × _____ square yards = _____ square yards

    ___

    STEP 2 Find the area of the whole squares:

    _____ × _____ = _____ squares

    _____ squares × _____ square yards = _____ square yards

    ___

    STEP 3 Find the total area: _____ + _____ = _____ square yards

1 square = 16 square yards

Find the area of each shaded shape. Write the area in square units.

2.

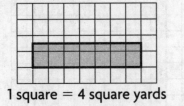

1 square = 4 square yards

___

3.

1 square = 9 square feet

___

4.

1 square = 4 square meters

___

On Your Own

Find the area of each shaded shape. Write the area in square units.

5.

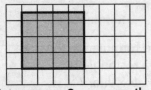

1 square = 9 square miles

___

6.

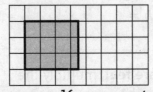

1 square = 16 square meters

___

7.

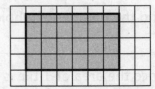

1 square = 25 square meters

___

Problem Solving REAL WORLD

8. A mosaic table top is shown. Each square has an area of 5 square inches. What is the area of the table top? **Explain.**

___

___

___

Table Top

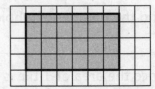

1 square = 5 square inches

Name _____

Multiply Three Factors

Essential Question How can you find the product of three factors?

UNLOCK the Problem REAL WORLD

You can use properties of multiplication to help make multiplication of three factors easier.

Sam ships 4 boxes of car model kits to Toy Mart. Each box contains 16 cartons, with 6 kits in each carton. How many car model kits does Sam ship?

- Underline what you are asked to find.
- Circle the numbers you will use to solve the problem.
- What operation can you use to solve the problem?

_____

Example Find 4 × (16 × 6).

STEP 1

Simplify the problem. Rewrite 4 × (16 × 6) as a product of two factors.

4 × (16 × 6) = 4 × (_____ × 16)   Commutative Property

= (4 × _____) × 16   Associative Property

= _____ × 16

So, 4 × (16 × 6) = 24 × 16.

So, Sam ships _____ car model kits.

STEP 2

Multiply.

$$\begin{array}{r} 16 \\ \times\ 24 \\ \hline \end{array}$$

☐ ← 4 × 16

+ ☐ ← 20 × 16

☐ ← Add.

Try This!

(18 × 8) × 3 = 18 × (_____ × _____)   Associative Property

= 18 × _____

= _____

Math Talk **Explain** how using properties makes it easier to multiple three factors.

Share and Show

1. Find the product of $7 \times (6 \times 13)$.

STEP 1 Simplify the problem.

Rewrite $7 \times (6 \times 13)$ as a product of two factors.

$7 \times (6 \times 13) = ($ _____ $\times$ _____ $) \times 13$

Associative Property

= _____ $\times$ _____

STEP 2 Multiply.

$$\begin{array}{r} 13 \\ \times\, 42 \\ \hline \end{array}$$

Find each product.

2. $3 \times (14 \times 3) =$ _____

3. $2 \times (4 \times 13) =$ _____

4. $(16 \times 6) \times 3 =$ _____

On Your Own

Find each product.

5. $7 \times (17 \times 4) =$ _____

6. $(18 \times 4) \times 6 =$ _____

7. $9 \times (17 \times 5) =$ _____

8. $(5 \times 26) \times 3 =$ _____

9. $9 \times (19 \times 2) =$ _____

10. $(21 \times 4) \times 6 =$ _____

Problem Solving REAL WORLD

11. There are 3 basketball leagues. Each league has 8 teams. Each team has 13 players. How many players are there in all 3 leagues?

_____

12. There are 8 boxes of tennis balls. There are 24 cans of tennis balls in each box. There are 3 tennis balls in each can. How many tennis balls are there in all?

_____

Name _____

Find Area of the Base

Essential Question How can you find the area of the base of a rectangular prism?

Connect The base of a rectangle is different than the base of a rectangular prism. The base of a rectangle is a side, but the base of a rectangular prism is a rectangle. To find the area of a rectangle, use the formula $A = b \times h$ or $l \times w$.

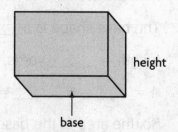

UNLOCK the Problem REAL WORLD

🔑 Example

Ana is making a diorama for a class project. The diorama is in the shape of a rectangular prism. She wants to paint the bottom of the diorama. What is the area of the base?

The base shape is a rectangle.
Use a formula to find the area.

$A = b \times h$

base = _____ inches

height = _____ inches

$A = \underline{\quad} \times \underline{\quad}$

$A = \underline{\quad}$ square inches

So, the area of the base of the

diorama is _____ square inches.

- What shape is the base of the diorama?

  _____

- What are the base and height of the base of the diorama?

  _____

5 in.

4 in.

11 in.

4 in.

11 in.

Math Talk Why would multiplying 11 by 5 give an incorrect answer for the area of the base?

Remember

Area of a rectangle:
$A = b \times h$ or $l \times w$

Area of a square: $A = s \times s$

Share and Show

1. Find the area of the base of the rectangular prism.

    The base shape is a _____.

    length = _____ yards, width = _____ yards

    $A =$ _____ × _____ = _____ square yards

    So, the area of the base is _____ square yards.

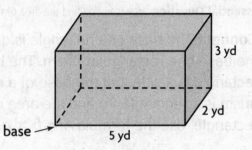

base →

3 yd

2 yd

5 yd

Find the area of the base of the rectangular prism.

2.

    3 in.

    2 in.

    2 in.

    _____

3.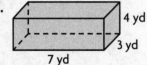

    4 yd

    3 yd

    7 yd

    _____

4.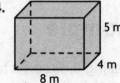

    5 m

    4 m

    8 m

    _____

On Your Own

Find the area of the base of the rectangular prism.

5.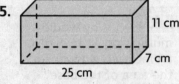

    11 cm

    7 cm

    25 cm

    _____

6.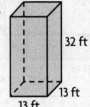

    32 ft

    13 ft

    13 ft

    _____

7.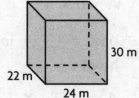

    30 m

    22 m

    24 m

    _____

Problem Solving REAL WORLD

8. Julio makes sugar cubes for horses. Each sugar cube edge is
    1 centimeter in length. He packs the sugar cubes in the box
    shown without gaps. Julio says he can fit 80 sugar cubes in the
    bottom layer. Is he correct? Explain.

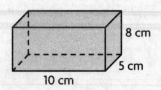

8 cm

5 cm

10 cm

_____

_____

_____

_____

Name _____

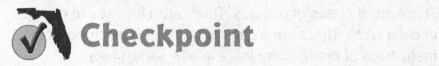

Concepts and Skills

Find each product. (pp. P301–P302)

1. $(13 \times 8) \times 5 =$ _____

2. $7 \times (12 \times 8) =$ _____

3. $4 \times (17 \times 3) =$ _____

Find the area of the shaded shape. Write the area in square units. (pp. P299–P300)

4.
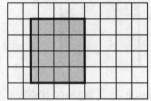

1 square = 4 square yards

_____

5.
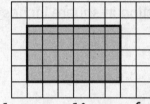

1 square = 16 square feet

_____

6.
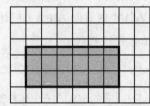

1 square = 25 square meters

_____

Find the area of the base of the rectangular prism. (pp. P303–P304)

7.
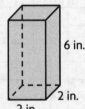

6 in.

2 in.
2 in.

_____

8.
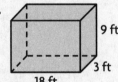

9 ft

3 ft

18 ft

_____

9.
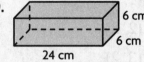

6 cm

6 cm

24 cm

_____

Problem Solving REAL WORLD

10. There are 6 grades competing in a spelling bee. Each grade has 10 teams. Each team has 4 members. How many members are competing in the spelling bee? (pp. P301–P302)

_____

Fill in the bubble completely to show your answer.

11. There are 9 crates of oranges. There are 18 boxes of oranges in each crate. There are 6 bags of oranges in each box. How many bags of oranges are there in all? **(pp. P301–P302)**

Ⓐ 108

Ⓑ 162

Ⓒ 972

Ⓓ 1152

12. A small tiled balcony is shown. Each tile is 9 square inches. What is the area of the shaded section in square inches?
(pp. P299–P300)

1 square = 9 square inches

Ⓐ 20 square inches

Ⓑ 144 square inches

Ⓒ 162 square inches

Ⓓ 180 square inches

13. Which ordered pair names point *A* on the grid? **(pp. P297–P298)**

Ⓐ (1, 5)

Ⓑ (2, 3)

Ⓒ (3, 2)

Ⓓ (5, 1)

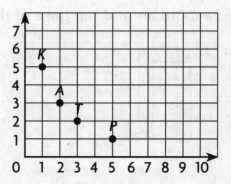

14. What is the area of the base of the rectangular prism?
(pp. P303–P304)

Ⓐ 40 square meters

Ⓑ 48 square meters

Ⓒ 144 square meters

Ⓓ 432 square meters

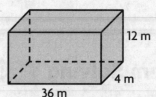

12 m

4 m

36 m

Comparative Relational Thinking

Use comparative relational thinking to tell whether the equation is *true* or *false*.

1. $61 - 57 = 53 - 49$

Think: How does 53 relate to 61? How does 49 relate to 57?

_____**true**_____

2. $97 + 33 = 88 + 24$

_____

3. $74 - 17 = 67 - 10$

_____

4. $28 + 76 = 35 + 69$

_____

5. $45 + 58 = 53 + 50$

_____

6. $36 - 27 = 47 - 16$

_____

Use comparative relational thinking to find the unknown number.

7. $55 + 84 = n + 76$

$n =$ _____

8. $91 - 64 = 103 - n$

$n =$ _____

9. $44 - 16 = n - 21$

$n =$ _____

Problem Solving REAL WORLD

10. Taylor ran 57 minutes on Tuesday and 48 minutes on Saturday. Mark ran 62 minutes on Tuesday and 53 minutes on Saturday. Taylor said they ran the same total time and Mark said they didn't. Who is right? Explain using comparative relational thinking.

_____

_____

_____

11. Mr. Paul drove from Drake to Clay and back. From Drake, he drove 312 miles, took a break, and then drove 244 miles to Clay. On his trip back to Drake from Clay, he drove 305 miles before he took a break. If he drove the same route back, how many miles did he drive the second part of his return trip?

_____

_____

_____

Lesson Check

1. Use comparative relational thinking to choose which equation is true.

   Ⓐ $56 + 42 = 63 + 49$

   Ⓑ $87 + 55 = 91 + 51$

   Ⓒ $61 - 37 = 57 - 41$

   Ⓓ $77 - 39 = 85 - 31$

2. A chef wants to use 15 pounds of red beans and 27 pounds of black beans for his chili. He only has 9 pounds of red beans and plenty of black beans. If he wants to maintain the total weight of beans, how many pounds of black beans should he use?

   Ⓐ 21 pounds

   Ⓑ 25 pounds

   Ⓒ 33 pounds

   Ⓓ 41 pounds

Spiral Review

3. Which number is 199,602 rounded to the place value of the underlined digit?
   (Lesson 1.4)

   Ⓐ 190,000

   Ⓑ 198,000

   Ⓒ 199,600

   Ⓓ 200,000

4. At the Sunset Center, a play ran for several weeks. In all, 36,284 people saw the play. What is the value of the digit 6 in 36,284? (Lesson 1.1)

   Ⓐ   6

   Ⓑ   60

   Ⓒ   6,000

   Ⓓ 60,000

5. In May, 138,075 people attended a sporting event. In June, 83,805 people attended a sporting event. How many more people attended the sporting event in May than in June? (Lesson 1.3)

   Ⓐ   54,270

   Ⓑ   55,870

   Ⓒ   58,770

   Ⓓ 155,870

6. Todd read the number "three hundred twelve thousand, sixty-two" in a newspaper article. What is this number in standard form? (Lesson 1.2)

   Ⓐ 302,620

   Ⓑ 320,062

   Ⓒ 312,062

   Ⓓ 312,620